Quality Management
of Nutraceuticals

ACS SYMPOSIUM SERIES **803**

Quality Management of Nutraceuticals

Chi-Tang Ho, Editor
Rutgers University

Qun Yi Zheng, Editor
Pure World Botanicals.

American Chemical Society, Washington, DC

Library of Congress Cataloging-in-Publication Data

Quality management of Nutraceuticals / Chi-Tang Ho, Qun Yi Zheng, [editors].

p. cm.—(ACS symposium series ; 803)

Includes bibliographical references and index.

ISBN 0–8412–3773–5

1. Functional foods—Analysis—Congresses. 2. Dietary supplements—Analysis—Congresses.

I. Ho, Chi-Tang, 1944- II. Zheng, Qun Yi, 1957- III. Series.

QP144.F85 Q35   2001
613.2—dc21   2001046390

Foreword

The ACS Symposium Series was first published in 1974 to provide a mechanism for publishing symposia quickly in book form. The purpose of the series is to publish timely, comprehensive books developed from ACS sponsored symposia based on current scientific research. Occasionally, books are developed from symposia sponsored by other organizations when the topic is of keen interest to the chemistry audience.

Before agreeing to publish a book, the proposed table of contents is reviewed for appropriate and comprehensive coverage and for interest to the audience. Some papers may be excluded to better focus the book; others may be added to provide comprehensiveness. When appropriate, overview or introductory chapters are added. Drafts of chapters are peer-reviewed prior to final acceptance or rejection, and manuscripts are prepared in camera-ready format.

As a rule, only original research papers and original review papers are included in the volumes. Verbatim reproductions of previously published papers are not accepted.

ACS Books Department

Contents

Bioactivity of Nutraceuticals

Indexes

Preface

The nutraceutical industry is a fast growing industry. The global nutraceutical and dietary supplement market reached $46 billion in 1999. However, for the adequate and healthy growth of this industry, it is necessary to properly communicate with consumers the good science most of the nutraceutical companies use to guarantee the chemical quality and biological efficacy of their products.

The symposium upon which this book is based was developed to bring together the leading scientists from industry, academia, and government to discuss and share information about the quality management of nutraceutical products. It is based on the two day symposium: *Quality Management of Nutraceuticals* that was part of the 220th National American Chemical Society (ACS) meeting in Washington, D.C., August 2000.

This book has been arranged into three sections. An introductory section introduces the topic of nutraceuticals and the analysis of the most commonly occurred phytochemicals in nutraceutical products, flavonoids, and anthocyanins. The next section covers the analysis of specific nutraceuticals such as black cohosh, saw palmatto, noni, cranberry, tea, and cocoa. The last section reviews the bioactivity of several nutraceutical products.

Acknowledgments

The editors acknowledge with great appreciation the financial assistance from the following sponsors: ACS Division of Agricultural and Food Chemistry, ACS Corporation Associates, Pure World Botanicals, Inc., and Nature's Herbs.

Chi-Tang Ho
Department of Food Science
Rutgers University
65 Dudley Road
New Brunswick, NJ 08901

Qun Yi Zheng
Pure World Botanicals, Inc.
375 Huyler Street
South Hackensack, NJ 07606

Introduction and Overview

Chapter 1

Quality Management of Nutraceuticals: Intelligent Product-Delivery Systems and Safety through Traceability

Paul A. Lachance and Raymond G. Saba

Department of Food Science, Rutgers, The State University of New Jersey, 65 Dudley Road, New Brunswick, NJ 08901–8520

Nutraceuticals are naturally derived, bioactive (usually phytochemical) compounds that have health benefits. The nutraceutical consumer product may be delivered as a dietary supplement and/or as a functional food. Functional food products, either naturally occurring or fortified, containing nutraceuticals must be managed to assure quality from "farm gate to plate." Assurance of nutraceutical quality requires a systems approach, commencing with the optimization of the agriceutical crop. Intelligent product-delivery systems philosophy requires identity of the crop botanically, with farm-site identification (via Global Positioning Systems), biodata records, agronomic practice records (Good Agricultural Practices). Phytochemical fingerprinting data requires analyses and tracking of the bioactive nutraceutical components and/or biomarkers. This knowledge requires the development of rapid assay methods and tools to assure real time compliance and standardization with documentation and monitoring for changes during handling, harvesting, processing and production (manufacture into product). Rapid assays can also be used to assure microbiological and chemical safety throughout the shelf-life of the product. Issues currently limiting full quality management need to be established and implemented before out-dated approaches become policy by default.

The passage of the 1994 Dietary Supplement Health and Education Act (DSHEA) stimulated a number of larger purveyors of vitamin/mineral supplements to enter or increase their market presence in the field of herbals. Partnerships and acquisitions occurred as the market mushroomed and shifted from primary presence in health food stores to expanded presence in pharmacies and discount stores. It took a while to fill-up the pipeline into the medicine cabinets of consumers who purchased these natural remedies with structure function claims! The U. S. market for dietary supplements moved from 8 Billion dollars in 1993 to 14 Billion dollars in 1999.

However, the market flattened or "has gone South." Consumer confidence in dietary supplements is eroding and the number of critics is increasing and they are gaining media attention. One of the major criticisms is that the products are of questionable quality, especially since pre-market safety testing and FDA pre-approval are not required, and thus value to the consumer becomes an issue. Further the consumer is becoming more educated in this area and is wary of the offering of products, purportedly "standardized" to provide specific quantities of a biomarker and associated with certain functional expectancies, but delivered in widely different market forms under the same product name.

The Council for Responsible Nutrition (CRN) believes the industry must address these and related concerns by a proactive, coordinated and cohesive plan that (a) establishes a credible mechanism to ensure safety and quality of all dietary supplements; (b) places all product value information and claims on sound science; (c) implements the educational component of DSHEA with consumer and professional programs; and (d) expands the science base with new and collaborative research efforts. This symposium, aimed at adding new science to the monitoring of quality, offers an opportunity to introduce the value of the application of intelligent product-delivery systems to ensure and sustain quality.

Definitions

NUTRACEUTICALS are naturally-derived, bioactive (usually phytochemical) compounds that have health promoting, disease preventing or medicinal properties. Nutraceutical compounds or substances can be delivered in the form of food (functional food) or as a dietary supplement, or in both forms (*1-3*).

QUALITY (of a product) is a measure of how closely the attributes of the product compare to established standards (however the standards may be emerging and therefore can be periodically upgraded). Product quality is measured, in hard terms, by the data that support the product attributes.

As shown in Figure 1, the overall and final value of the product needs the input of safety and stability, etc. to assure product efficacy and therefore product quality. However the quality of a complex product hinges on the quality of each ingredient and, their formulation, and it is limited by the lowest quality provided by the supplier(s), and therefore the need for supplier quality management at the earliest source level. Therefore, QUALITY MANAGEMENT of a product requires an integrated set of procedures or practices, as to specific description, agronomic or equivalent practices, harvesting techniques, holding and manufacturing procedures, etc, that are developed and implemented for the purpose of assuring a level of product quality (efficacy, stability and safety).

QUALITY MANAGEMENT is needed. Defining a product in terms of consistent quality is the basis for establishing final product value. Generating reliable scientific data is a must but the data must be based upon sustaining the assurances of safety and efficacy throughout the stages involved in the source and stages in the progress to the final product and thus product value. The conduct of clinical trials on poorly defined complex botanical materials provides highly questionable data.

Issues which drive the demand for a quality management mechanism are public, technological and financial.

The public issues are issues such as (a) the increased global sourcing of nutraceutical/botanical/food/ingredients; (b) the increased health concerns of baby boomers; (c) the increase in government funding of safety concerns.

The technological issues are: (a) Memory capacity and miniaturization, information sharing systems, (b) "smart" devices (information gathering/sharing), (c) dynamic (read/write) mediums, (c) Internet capabilities, (etc).

The financial issues relate to marketing "narcissism" and the desire of rapid return on investment with minimal costs for implementing and maintaining quality management.

The challenges pertain to (a) available technology; (b) the benefits to the industry of standardization; (c) decisions concerning quantitative research needs; (d) regulatory "paranoia", (e) education variability; (f) claims compliance;

Incorporating traceability makes possible the standardization of product quality and value.

Intelligent Product-Delivery Systems

IPDS is a program (*4-5*) that uses information technology to deliver greater product value throughout the resource/product supply chain.

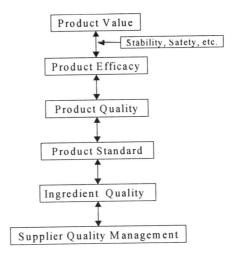

Figure 1. A quality management infrastructure is needed to assure product value.

Traceability permits and assures a degree of scientific underpinning vital in this era of global sourcing. Standardization to date has been primarily based upon botanical and biomarker information with little reference to the precise source and conditions at that source and throughout the "travel" of the article of trade. The significant quality attributes assuring safety (microbiological, chemical and physical), as well as quality assuring efficacy of the product (or its constituents) can be optimized by a resource traceability system (RTS) as part of an Intelligent Product-Delivery System (IPDS). With such systems in place from field (farm gate) to plate and consumer disposal, the purveyor provides extensive accountability and substantially increases the probability that the product will be in compliance with preset standards. Further, if a recall becomes necessary, it can be rapidly and effectively executed. "Accidents" are thwarted, and death, as a negative variable, is avoided.

As shown in Figure 2, IPDS provides product tracking and therefore traceability by gathering information (I.D. tags in the form of chips, sensors, bar codes, portable data files, etc.) from source to disposal. Between agriculture and consumption, IPDS makes possible rapid traceability through I.D. tags with portable data and tabulation of related information describing the origin, harvest and history of the product.

6

Food Traceability: Information links through I.D. tags (portable data files, intelligent bar codes, etc.) for rapid traceability of food in both directions between agricultural source and consumption.

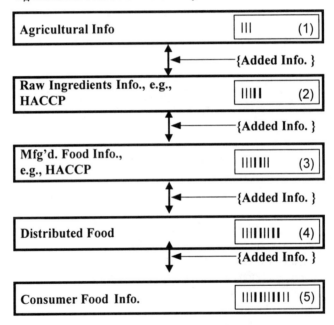

Figure 2. Food traceability information linkages in an Intelligent Product-Delivery System.

As shown in Table I, to assure quality in terms of both safety and efficacy, a series of key standards for resource traceability are needed:

Table I. Standardization Needs of a Quality Management Plan.

QUALITY MANAGEMENT
TO ASSURE SAFETY & QUALITY, STANDARDIZATION MUST BE ESTABLISHED:
(a) Materials defined by Latin name (or) chemical entity w/CAS# or chemical source/synthesis location
(b) GIS descriptors of plant material
(c) Sample verification of plant material by systematic botanist or chemical certification
(d) Certification of below tolerance levels of: - organics (e.g., pesticide residues) - heavy metals (e.g., lead, mercury) - microbiological pathogens, etc.
(e) If known, content of bioactive(s)/biomarker(s)
(f) Documentation of GMP; institution of HACCP and monitoring of FDA database for adverse effects

First, the material (e.g., plant) must be defined by the Latin descriptor (or) the chemical entity defined in Chemical Abstract System nomenclature.

Secondly, a descriptor of the place of origin (e.g., field, or a manufacturer's facility for a synthesized entity) must be provided. For a plant material, use of a Global Information System (GIS), as shown in Figure 3, provides the latitude and longitude map of the field in which the plant cultivar was seeded, cultivated or harvested. The GIS information can now be linked to weather history at this location with identification of the cultivar and data on the applied fertilizers, pesticides, etc.

Thirdly, it is imperative that a sample of the plant material be verified botanically or that the chemical entity commodity be certified.

Fourthly, certification must exist that the parameters that assure biochemical and microbiological safety are below national/international tolerance levels. This applies to relevant microbiological pathogens, heavy metals and organic chemical residues (e.g., pesticides). Bio and chemical assays are under development permitting rapid and repeated monitoring for safety and efficacy.

Fifthly, if known, the bioactive and /or biomarker content and concentration can be recorded.

8

Fifthly, if known, the bioactive and /or biomarker content and concentration can be recorded.

Lastly, there must be documentation of Good Manufacturing Practices (GMP), compliance with the principles of Hazard Analysis and Critical Control Points (HACCP) and linkage to FDA, or similar international databases, for monitoring adverse effects that may be reported.

GIS Application

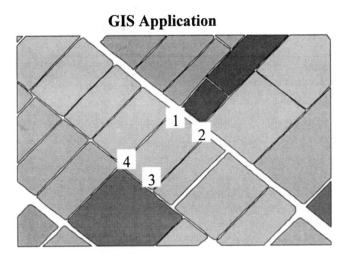

Figure 3. A global Information System (GIS) representation of an agriceutical crop. Fields are represented by polygons delimited by black lines. Fields of the same shade are planted with the same cultivar . Information can be linked to different farms, fields within farms, or even to regions within fields. Growers apply pesticides and fertilizers on a field-by-field basis and this information is easily mapped. In a GIS map each point (pixel) (1,2,3 and 4) has an accurate location (latitude & longitude) reference (6). The coordinate system used is UTM NAD 83.

In this model of Quality Management, IPDS/RTS reveals and documents:
Agricultural data
Raw ingredient data
Manufacturing data
Distribution data

In summary, the state-of-the-art of the emerging nutraceuticals industry requires an emphasis on novel approaches to quality management. IPDS, with RTS, provides a unique and dynamic approach that can help to deliver the promise that nutraceuticals can provide: major reductions in the pathogenesis of chronic disease and in the associated costs of health care; and thus, a better quality of life.

Acknowledgement

A contribution of the Nurtraceuticals Institute, a joint program of the New Jersey Agricultural Experiment Station, Rutgers University, and the Center for Food Marketing, St. Joseph's University.

References

1. Guhr, G.; Lachance, P. A. In *Nutraceuticals: Designer Foods III Garlic, Soy and Licorice*; Lachance, P. A., Ed.; Food & Nutrition Press, Inc.: Trumbull, CT, 1997, pp. 310-364.
2. Dillard, C. J.; German, J. B. *J. Sci. Food Agric.* **2000**, *80: 1744-1756.*
3. *Essentials of Functional Foods;* Schmidl, M. K.; Labuza, T. P., Eds.; Aspen Publishers, Inc.: Gaithersburg, MD, 2000; pp 395.
4. Yam, K. L.; Saba, R. G*., Packaging Technology & Engineering*, **1998**, *7* (3), pp. 22-26.
5. Yam, K. L.; Saba, R. G.; Lachance, P. A.; Delprat, J., U.S. Patent Application 09/099,862, 1998.
6. Oudeman, P. Rutgers University, personal communication.

Chapter 2

The 6S™ Quality Management of Nutraceuticals: An Operating Principle at Pharmanex

Andrew Chiu[1], Joe Chang[1], and Michael Chang[2]

[1]Pharmanex, Inc., 75 West Center Street, Provo, UT 84601
[2]Pharmanex, Inc., 2000 Sierra Point Parkway, Brisbane, CA 94005

We describe the evidence-based 6S™ Quality Management Process that has been implemented by Pharmanex to ensure the production of the highest quality neutraceutical products for consumers. The 6S™ Process includes 6 key research elements, Selection, Sourcing, Structure analyses, Standardization, Safety and Substantiation, that govern all the steps from R&D to clinical testing and to product marketing. Key features of this process are chemical characterization of the botanical extract, identification of the biologically active components, standardization of extracts in terms of their chemical composition and bioactivity, animal toxicity studies, and controlled clinical trials to rigorously assess efficacy and safety. The development by Pharmanex of the dietary supplement Cholestin™, which maintains healthy cholesterol levels, is discussed as an example of the use of the 6S Process to produce a scientifically validated product with batch-to-batch consistency.

Sales of botanical products are approaching $3.5 billion annually and 60 million American adults use herbs in their efforts to promote health. This growth in interest and the use of traditional therapies has been fueled in part by the desire of aging baby boomers to find alternatives to pharmacological drugs and by an increase in consumer awareness of the potential benefits of botanical products.

Nevertheless, despite their popularity, the health claims of traditional remedies are in many cases not based on sound research into their mechanisms of action, toxicity, and clinical efficacy. Such a situation does not conform to good pharmacological practices and would not be tolerated by healthcare consumers if it was the standard for pharmaceutical drugs. It has also given rise to the common but unsubstantiated assumptions that all herbal products are safe since they are natural, have no side effects, and are efficacious over a broad dose range. However, the lack of rigorous scientific research for the majority of botanical products must be rectified if the reluctance of many in mainstream medicine to accept these products as medically valid is to be overcome.

Regulatory Issues

Botanicals are currently regulated as dietary supplements under the Dietary Supplement Health and Education Act (DSHEA) and are defined as plant extracts, enzymes, vitamins, and hormonal products that are available to the consumer without prescription. They are limited to "structure/function" claims - that a product may affect the structure or function of the body. Since dietary supplements are not required to meet the same regulatory requirements as pharmaceutical drugs, explicit drug claims are not allowed in their labeling. This means that dietary supplements, in contrast to pharmaceutical drugs, cannot be marketed for the "treatment, diagnosis, cure or prevention of a disease".

One factor that may drive producers of botanicals towards generating a more scientific knowledge base for traditional herbal products may be the fact that research methodologies paralleling those used with pharmaceutical drugs will have to be used if herbal products are to be approved by the Food and Drug Administration (FDA) as botanical drugs. The FDA has recently issued draft procedures for obtaining marketing approval for a botanical drug through an over-the-counter (OTC) monograph or a New Drug Application (NDA) *(1)*. To be able to sell a botanical as a drug under an existing OTC monograph, a product would have to be generally recognized as safe and effective, and there would have to be existing clinical studies that confirmed the product's safety and effectiveness. Botanical products not meeting these criteria would have to be approved using an NDA, which would need to include clinical studies, safety data and chemistry, manufacturing, and controls (CMC) information. If existing information was not sufficient to support an NDA, an Investigational New Drug (IND) application for the product would have to be submitted. Although the amount of data that would have to be supplied for the NDA would vary depending on whether the product has been marketed in the US, outside the US or has not been marketed, considerable CMC and toxicological information might be required, along with phase I, phase II, and phase III clinical trials. Whatever the outcome of the debate over these guidelines, it appears as though regulatory approval for a disease claim for a botanical drug will require scientifically substantiated evidence similar to that required for a pharmaceutical drug.

Components of the 6S Process

If herbal extracts are to become a respected component of mainstream medicine, even though a disease claim may not be sought, we at Pharmanex believe that their composition and pharmacological properties should be rigorously defined. For example, the composition of preparations of the same herbal extract often differ from manufacturer to manufacturer and not even a single company may produce a standardized extract. This makes it difficult to compare the efficacy and safety of these extracts either in the literature or in new clinical trials. Further, the efficacy of botanicals is often based on anecdotal data and a long history of use, which is not considered satisfactory for pharmaceutical drugs. We have addressed such issues by developing the 6S™ Quality Management Process (Table I), which we believe offers a mechanism by which herbal products can be shown to be truly safe and efficacious and can be produced with a consistent pharmacological activity from different batches of raw materials.

We believe that the 6S Process addresses the fundamental problems of quality control that are found throughout the natural products industry and allow Pharmanex to offer the consumer the highest quality natural healthcare products available.

Chemistry Components of the 6S™ Process

Natural products can be subject to variation in composition, to contamination and to deterioration, so each batch should therefore be tested for both chemical composition and biological activity. Important factors that contribute towards batch consistency include an accurate taxonomic classification of the plant species, consistent use of only the part of the plant

Table I. The Components of the 6S™ Quality Management Process

Selection	*Standardization*
Review of scientific data to identify unique natural products	Consistent chemical profile
Selection of product for development base on unmet health need	Consistent pharmacological profile ensures consistent usage level
Sourcing	*Safety*
Raw material assessed for active or marker components.	History of safe use at recommended dosages
Quality control during harvest and storage	Toxin and animal safety analyses
Structure	*Substantiation*
Identification of active constituents	Efficacy claims based on substantiated scientific claims
Active constituents assayed by state-of-the-art validated methods	Prospective clinical study performed if necessary

that contains the pharmacolgically active compounds, definition of a chemically standardized extract, the absence of toxic contaminants, and the stability of the final product. Where possible, extracts should be standardized with respect to the content of the presumed bioactive components and these active constituents should be assessed by methods approved by the Methods Validation Program.

Pharmacology Features of the 6S™ Process

The pharmacological activity of a natural product is very often believed to be the result of the combined actions of several of its constituents. However, in many cases the active constituents of a complex herbal extract have not been completely characterized. Chemical characterization of the product may thus not be a complete indicator of its pharmacological activity. We believe that a bioassay employing a clinically relevant activity is a better way to ensure consistency between batches of natural products with respect to the content of their bioactive constituents. This is why a key feature of the 6S approach is the investigation of the mechanism of action of the extract to identify both a molecular target for the bioassay and a surrogate molecular marker, the levels of which can be assessed as an endpoint in clinical studies.

The bioavailability of active constituents is another factor that can lead to differences in efficacy between different preparations of the same herb. Indeed, low bioavailability could lead to the absence of a pharmacological effect with an extract with the same composition of active ingredients as a comparator. Again, bioassays of active components *in vivo* may be more appropriate than chemical analysis. Such studies are central to the determination of optimal dose and dosing regimens.

Clinical Features of the 6S™ Process

Clinical research on herbal products has unfortunately often been lacking in the well-controlled clinical trials that are required by regulatory agencies evaluating conventional drugs, raising doubts about the validity of the clinical efficacy and safety of botanicals. Nevertheless, there exists the potentially dangerous assumption, especially in the absence of standardized preparations of a given herb, that botanical products are implicitly safe because they are natural products and have a long history of use. There is currently no systematic collection of data on the safety of herbal products, which has probably lead to an underreporting of adverse events and supports the general belief that herbal products have few side effects. At Pharmanex, we believe in only making efficacy claims that are supported by well-documented scientific studies and are committed to carrying out new clinical trials to support the health claims of our products. Further, we believe that adverse event data should be gathered in a rigorous, open-label safety study.

The Pharmanex clinical development strategy for a product therefore includes human bioavailability studies to select a dose, an initial exploratory study using a surrogate marker of biological activity, a double-blind, placebo-controlled clinical study to assess efficacy using this surrogate marker, and a multicenter, open-label study to assess safety. The next section illustrates how the 6S™ Process has been applied to the Pharmanex product: Cholestin™, a red yeast rice extract that promotes healthy cholesterol levels.

Application of the 6S™ Process to the Development of Cholestin™

Cholesterol is a component of cell membranes and a precursor of steroid hormones that is essential for cell viability in higher animals. The body's total cholesterol level is obtained from the diet and also from that synthesized *de novo* in the body. Cholesterol is transported through the circulation on different types of lipoprotein particles such as low-density lipoproteins (LDL), high-density lipoproteins (HDL) and very low density lipoproteins (vLDL) It has been long determined that a high total cholesterol level is a risk factor for cardiovascular disease. Elevated LDL-cholesterol is also a risk factor for cardiovascular disease because it promotes the formation of atherosclerotic plaque in arteries throughout the body. Oxidized-LDL particles, formed by the oxidation of circulating LDL particles by free radicals, can injure the arterial wall, stimulate the proliferation of smooth muscle cells and white cells near the inner surface of the arterial wall and induce the formation of foam cells that are the major cause of atherosclerotic lesion formation. Lowering total and LDL-cholesterol levels thus reduces the risk of coronary artery disease.

The rate-limiting step in cholesterol synthesis is the enzyme 3-hydroxy-3-methylglutaryl-CoA reductase (HMG-CoA reductase) and in 1979, Endo *(2)* discovered that a strain of *Monascus* yeast produced an inhibitor of cholesterol synthesis (monacolin K) that inhibited this enzyme. This metabolite was also isolated from *Aspergillus* and came to be known as lovastatin *(3)*. Cholestin is a red yeast rice extract that contains monacolins and has been shown to lower cholesterol levels, and has been developed and marketed by Pharmanex to enable people to supports the maintenance of healthy cholesterol levels.

Selection and Sourcing

Red yeast rice is prepared by fermenting rice with red yeast (*Monascus purpureus*) and its use in China was first described in the Tang Dynasty in 800 AD. In the ancient Chinese pharmacopoeia *Ben Cao Gang Mu-Dan Shi Bu Yi*, published in 1578, red yeast rice is recommended for improving the blood circulation *(4)*. In China today, red yeast rice is part of the daily diet and is used to make rice wine and for food coloring and flavoring. In the West, a combination of knowledge of traditional Chinese medicine with modern biochemistry and pharmacology has resulted in the development of Cholestin as an effective dietary supplement to promote normal cholesterol levels. It is also less expensive than pharmaceutical drugs marketed for the same purpose. Application of the 6S Process ensures that this Pharmanex product has been rigorously characterized in terms of its pharmacology, long-term toxicity, and in clinical trials of efficacy and safety (Figure 1). Furthermore, Cholestin is prepared by a proprietary process that ensures each batch is standardized with respect to the content of its biologically active components.

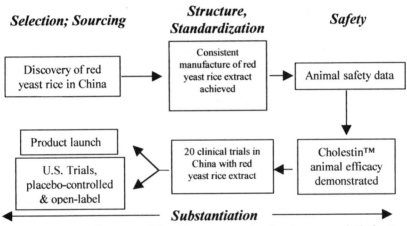

Figure 1. Application of the 6S process to the development of Cholestin.

Structure and Standardization

Red yeast rice also contains other monacolin-related substances that inhibit HMG-CoA reductase, such as sterols (β-sitosterol, camesterol, stigmasterol, and sapogenin). Isoflavones and isoflavone glycosides, and monounsaturated fatty acids are also found. These discoveries lead Chinese scientists in the 1980's to study the benefits of red yeast rice on the cardiovascular system, and in 1993, a proprietary manufacturing process was developed for a standardized red yeast rice product which later became known as Cholestin, which reproduced the beneficial properties of the traditional Chinese extract.

Cholestin has been extensively characterized by Pharmanex and its composition is shown in Table II.

Table II. Composition of Cholestin™ Red Yeast Rice Extract (6)

Component	Percentage by Weight
Total sugars	73.4
Fiber	0.8
Protein	14.7
Moisture	6
Total natural pigment	0.3
Ash	2.45
Phosphorus	0.4
Trace elements[a]	Trace
Total HMG-CoA[b] reductase inhibitors (10 Monacolins)	0.4
Fatty acids	
Saturated (palmitic and stearic)	<1.5
Mono- and polyunsaturated, e.g., oleic, linoleic, linolenic	<1.5

[a]Calcium, aluminum, iron, manganese, magnesium, copper, silver
[b]HMG-CoA, 3-hydroxy-3 methylglutaryl coenzyme A

The Pharmanex proprietary manufacturing process produces consistent batches of red yeast rice that are standardized to contain 0.4% total HMG-CoA reductase inhibitors. Thus, there is about 9.6 mg of HMG-CoA reductase inhibitors in the recommended daily dose of 2.4 g of Cholestin, of which 5 mg is lovastatin. It should be noted that comparison of the cholesterol-lowering effect of Cholestin with that of lovastatin suggests that this dose of Cholestin is equivalent to 20 mg of lovastatin (6). Thus, the inhibition of cholesterol biosynthesis by Cholestin cannot be explained by its lovastatin content alone but rather by a combination of the actions of all the monacolins and other compounds in the red yeast rice extract. As described below, Cholestin treatment also lowers serum triglyceride levels, which may be due to the unsaturated fatty acids found in the extract. Thus, the mechanism of action of Cholestin is at least partially understood, allowing cholesterol levels to be used as a valid surrogate marker for the effects of Cholestin in reducing blood lipid levels and mitigating cardiovascular risk factors.

To examine the proposed mechanism of action of Cholestin, several studies have been performed on its effects on blood lipid and lipoprotein levels in animal models of hypercholesterolemia. In one, serum total cholesterol and LDL-cholesterol levels were reduced by 45% - 60% in rabbits fed Cholestin for 30 days plus a diet that induced endogenous hypercholesterolemia, compared to animals receiving the diet alone. Similar decreases were also found in rabbits with exogenous hyperlipidemia that were treated with Cholestin. In addition, lesions in the aortas and lipidosis in the livers of Cholestin-treated rabbits were less severe than those in control animals. Similar data were obtained in quail fed an atherogenic diet. In each of these animal models, Cholestin treatment elicited changes in serum lipid levels similar to those seen with mevinolin (7).

Safety

Toxicology studies with doses of Cholestin many times greater than the recommended daily human dose have shown little or no toxicity in both long and short-term treatment. In an acute toxicity study, a *Monascus purpureus* (red yeast) extract showed minimal adverse effects in mice after administration of 16g/kilogram body weight, a dose 533 times a typical human dose, with no deaths (8). In a long-term study, rats showed no significant evidence of toxicity over 90 days of daily administration of doses of Cholestin up to 125 times the normal human dose, including no effects on liver or kidney function. Cholestin was shown to be negative in an Ames bacterial mutagenicity assay, and to not damage either mouse bone marrow cells or sperm (9).

Substantiation

It is a central principle of the Pharmanex 6S Process that, if necessary, additional clinical trials will be conducted to confirm existing data and to rigorously demonstrate the efficacy and safety of a product. This is the case with Cholestin. In the United States, two published double-blind, placebo-controlled studies have shown the efficacy of Cholestin (10, 11). In China, which used a red yeast rice product produced by the same process as Cholestin, of over 30 studies translated into English, approximately half of these were controlled, including 3 double-blind, placebo-controlled trials and a number of trials using "positve" (active) controls. In the Chinese studies, patients with moderate or severe hyperlipidemia were treated with the Cholestin-like product with doses of 0.6 to 1.2 g.day for 8 weeks. Overall, the following changes in serum lipids were obtained: total serum cholesterol, 11.3 to 27.5% decrease; serum triglycerides, 18.6 to 43.0% decrease; LDL-cholesterol, 22.4 to 34.2% decrease; HDL-cholesterol, 5.0 to 26.4% increase.

We therefore decided to conduct a double-blind, placebo-controlled, prospectively randomized 12 week controlled trial at a US university research center. Eighty-three hyperlipidemic adults were placed on a diet similar to the American Heart Association Step 1 diet for 1 week and then were randomly assigned to receive either rice powder placebo or 2.4 g Cholestin capsules daily for 12 weeks. It was found that in the Cholestin-treated group that total cholesterol had decreased significantly by 16% compared to placebo by week 12 and that LDL-cholesterol and total triglycerides were also reduced significantly, by 23% and 15%, respectively. There was no significant change in HDL-cholesterol. No serious adverse events were reported in this study and liver function tests were not affected by Cholestin treatment (10). This study thus confirmed the results of the Chinese studies in a typical US patient population consuming a western diet.

A larger, multi-center, open-label study has also been carried out in the US (11). In this study, 187 patients with mild to moderate hypercholesterolemia were placed on the AHA Step 1 diet for 4 weeks. They were then given Cholestin (1.2 g twice a day) for 8 weeks. It was found that maintenance on the Step 1 for 4 weeks alone had no effect on serum cholesterol. However, after 8 weeks of Cholestin treatment, total cholesterol was reduced by 16.4%, LDL-cholesterol was reduced by 21.0%, triglycerides were reduced by 24.5% and HDL-cholesterol was increased by 14.6%. When Cholestin treatment was discontinued for 2 weeks, serum lipid levels returned to pre-treatment levels. Treatment-related adverse events were reported by 18% of patients and included headache, abdominal bloating and gas; however, this study judged Cholestin to be well-tolerated.

Conclusions

Application of the Pharmanex 6S Process to the development of Cholestin has resulted in a dietary supplement that has been scientifically demonstrated to be safe and efficacious. The biologically active constituents of the extract have been characterized and standardized extracts are produced with consistent pharmacological activity from batch to batch. We believe that such a process is essential across the industry if the true value of herbal products for supplemental and medicinal use is to be established.

Acknowledgements

The authors wish to acknowledge the following individuals for their contributions to the operational practices of the 6S™ quality process: Drs. R. Cooper, C. Smidt, W. Duersch, H. Sun, J-s. Zhu, and D.C. Zhang. We also wish to thank Dr. D. Burke for his editorial contribution to this manuscript.

References

1. Botanical Drug Products. Draft Guidance for Industry. FDA. *Federal Register*, August 11, 2000.
2. Endo, A. *J. Antibiot.* **1979**, *32*, 852-854.
3. Tobert, J.A.; Bell, G.D.; Birtwell, J.; James, I.; Kukuvetz, W.R.; Pryor, J.S.; Buntinx, A.; Holmes, I.B.; Chao, Y-S.; Bolognese, J.A. *J. Clin. Invest.* **1982**, *69*, 913-919.
4. Ying Sing, S. In *T'ien Jung K'ai Wu-Chinese technology in the Seventeenth Century;* Sun E-T; Sun S-C, transl; London:Pennsylvania State University Press, 1966; pp.291-294.
5. Bradford, R.H.; Shear, C.L.; Chremos, A.N; Dujovne, C.A.; Franklin, F.A.; Grillo, R.B.; Higgins, J.; Langendorfer, A.; Nash, D.T.; Pool, J.L.; Schnaper, H. *Am. J. Cardiol.* **1994**, *74*, 667-673.
6. Ma, J.; Li, Y.; Ju, D.J.; Li, J.; Qing, Y.; Hua, Y.; Zhang, D.; Cooper, R.; Chang, M. J Agric Food Chem (submitted for publication, 2000).
7. Li, C.; Zhu,Y.; Wang,Y.; Zhu, J-S.; Chang, J.; Kritchevsky, D. *Nutr Res.* **1998**, *18*, 71-81.
8. Li, C.L.; Li, Y.F.; Hou, Z.L.; *Bull. Chinese Pharmacol. Soc.* **1995**, *12*, 3.
9. Liu, Z.; Guo, S.; Yao,X.; Liu, Z.; Shen, D.; Zhu, J-S.; Chang,J. Toxicity and mutagenicity of Cholestin ™, *Monascus purpureus* (red yeast) rice: A traditional food and herb of China. Pharmanex, Inc.:Brisbane, CA, 1999, pp.1-23 (unpublished).
10. Heber, D.; Yip, I.; Ashley, J.M.; Elashoff, D.A.; Elashoff, R.M.; Go, V.L.W. *Am. J. Clin. Nutr.* **1999**, *69*, 231-236.
11. Rippe, J.; Bonovich, K.; Colfer, H.; Davidson, M.; Dujovne, C.; Fried, D.; Greenspan, M.; King, S.; Karlsberg, R.; LaForce, C.; Litt, M.; McGhee, M.R. *Circulation* **1999**, *99*, 1123 (Manuscript submitted to Journal of General Internal Medicine for publication).

Chapter 3

Analysis of Flavonoids in Botanicals: A Review

Howard M. Merken[1,2] and Gary R. Beecher[1]

[1]Food Composition Laboratory, Beltsville Human Nutrition Research
Center, Agricultural Research Service, U.S. Department of Agriculture,
Beltsville, MD 20705
[2]Current address: Chemistry Department, 2339 Sciences Building,
Southern Illinois University at Edwardsville, Edwardsville, IL 62026–1652

Flavonoids, known for their antioxidant activities in foods, are found in some botanicals. High-performance liquid chromatography (HPLC) has become the method of choice for flavonoid analysis in foods and botanicals. Other chromatographic methods, such as countercurrent chromatography (CCC), high-performance thin-layer chromatography (HPTLC), micellar electrokinetic capillary chromatography (MECC), and thermospray-mass spectrometry (TSP-MS), have also been used. The flavonoids found have been chiefly the glycosides of flavones and flavonols, with some coumarins and flavan-3-ols. Three flavanones were identified in *Limonium sinense,* two flavanones were found after hydrolysis of *Mentha piperita,* and the flavanonol glycoside astilbin was found in St. John's wort. Isoflavones have been found in *Ononis spinosa* and *Sophora japonica.*

Flavonoids, diphenylpropanoids of low molecular weight found in vascular plants, are consumed by humans in the west at 100-1000 mg/day (*1, 2*). Most importantly, as antioxidants, they scavenge radicals by donating hydrogen, helping to prevent low density lipoproteins (LDL) from becoming oxidized, which could lead to coronary heart disease (*3*). Their beneficial health effects also include their ability to complex transition metal cations such as Cu^{2+} and Fe^{2+} (*3, 4*). Other activities include anti-AIDS, anti-arthritic, anticancer, anti-hypertensive, anti-inflammatory, and antiviral activity (*2*). Flavonoids regenerate α-tocopherol by reducing the α-tocopherol radical (*3*). Flavonoids are also believed to regenerate ascorbate, which regenerates Vitamin E (*5*).

Mabry and colleagues (*6*) published nmr (nuclear magnetic resonance) spectra of 128 flavonoids, and the ultraviolet-visible (UV-vis) spectra of 175 flavonoids, molecular extinction coefficients, and UV spectral data in different solvents. Daigle and Conkerton (*7, 8*) reviewed the analysis of flavonoids by HPLC. A review by Robards and Antolovich (*9*) concentrated on the analytical chemistry of fruit flavonoids. The HPLC systems of food flavonoids published from 1988 to early 1999 was reviewed by Merken and Beecher (*10*). Middleton (*2*) reviewed the biological properties of flavonoids. Rice-Evans and colleagues (*3*) explored the mechanistic reasoning behind the antioxidant activities of flavonoids. A wealth of information about the health affects of flavonoids in both foods and botanicals is found in a book edited by Rice-Evans and Packer (*11*).

Flavonoids are also found in various botanicals. Flavonoid analysis is thereby necessary for an understanding as to the safety and efficacy of botanical preparations (*12*). This review concentrates on analysis of plants and plant extracts, as opposed to analysis of commercial preparations of botanicals or analysis of flavonoids in human tissue, blood, urine, or feces.

The literature often neglects the common names of botanicals, which in this paper were often gleaned from various websites. The literature also varies in the nomenclature of relevant flavonoid glycosides and the coumaric esters of flavonol glycosides.

Botanical Analysis

Sample Preparation

Botanical samples were generally first treated by liquid-liquid extraction (LLE). A solid phase extraction (SPE), using for example a Sep-pak C_{18} column (*13*) or a Bond Elut C_{18} cartridge (*14*), sometimes followed.

Leaves of *Apocynum venetum* were dried, roasted twice, and extracted and partitioned in liquid phases (LLE) (*15*). The portion soluble in ethyl acetate (EtOAc) was put through a Sephadex LH-20 column. Fifteen compounds were isolated through reversed-phase preparative thin-layer chromatography (TLC).

Researchers often deglycosylated the flavonoids via acidic hydrolysis. In separate tests, Vilegas and colleagues (*16*) used 6% HCl and MeOH/2% HCl to deglycosylate two new flavonoid glycosides from *Maytenus aquifolium.* The monosaccharides from the methanolysis were later analyzed by GC-MS.

Analysis of Botanicals

HPLC conditions were similar to those used for analysis of flavonoids in foods (*10*). Reversed-phase (RP) columns were usually used, although a normal phase silica gel column was used for the proanthocyanidin-rich Pycnogenol (*17*). Flow rates ranged from 0.8-2.0 ml/min.

Detection was generally done using a diode-array detector (DAD), which can collect data over the interval of several hundred nm, and can produce chromatograms over several preset wavelengths. Detection was carried out from 254-370 nm. Anthocyanidins were generally not found in botanical leaves or flowers, obviating analysis at 465-560 nm (*10*). In analyzing procyanidins in flowers and leaves of *Crataegus* spp., Rohr and colleagues (*14*) used 220 nm DAD. In comparing electrochemical detection (ECD) to diode-array detection for procyanidins, they suggested DAD, which they found more selective and easier to handle, despite the greater sensitivity of ECD.

Other Analytical Methods

Non-volatile compounds can be analyzed quickly with liquid chromatography-mass spectrometry (LC-MS) (*18*). Thermospray-mass spectrometry (TSP-MS), using a buffer of ammonium acetate, was used to identify flavonol di- and triglycosides from the flowers of *Calendula officinalis* and the leaves of *Ginkgo biloba* and *Tilia cordata.* This "soft-ionization" technique limits the fragment ions so that the sugar and the aglycone can be determined.

Hahn-Deinstrop and Koch (*19*) used thin-layer chromatography (TLC) in the identification of terpene lactones and flavone glycosides in *Ginkgo biloba.* Evaluating high-performance thin-layer chromatography (HPTLC) plates densitometrically, Jamshidi and colleagues (*20*) claimed that HPTLC is as good as HPLC for analysis of pharmaceuticals. Maisenbacher and Kovar (*21*) used displacement chromatography to prepare St. John's wort oil. Their analytical methods included HPLC, HPTLC, and photometry.

Shi and Niki (*26*) list several reports of the health effects of *Ginkgo biloba* extract, which include the scavenging of hydroxyl and peroxyl radicals, the interaction with biologically occurring nitric oxide, and prevention of hydroxyl radical-induced cerebellar neuron death in rats. Akiba and colleagues (*35*) showed that GBE can suppress aggregation of platelets exposed to a combination of Fe^{2+} with either *tert*-butyl hydroperoxide or hydrogen peroxide, while oxidative stress induced platelet aggregation was not affected by ginkgolides A, B, and C.

Vanhaelen and Vanhaelen-Fastre (*22*) combined gradient elution countercurrent chromatography (CCC) and preparative HPLC to isolate and purify seven flavonol glycosides from the leaves of *Ginkgo biloba*. Yang and colleagues (*23*) used multidimensional counter-current chromatography (MDCCC) to separate flavone aglycones found in crude mixtures of *Ginkgo biloba* and *Hippophae rhamnoides*. Their system included two high-speed countercurrent chromatography (HSCCC) systems.

Micellar electrokinetic capillary chromatography (MECC) has higher column efficiencies than does HPLC (*24*). On-line UV detection can be used for detection in capillary electrophoresis (CE) as well as in HPLC (*25*). Analyzing the results at 260 nm in both cases, Pietta and colleagues (*25*) used MECC to separate several flavonol 3-*O*-glycosides found in *Ginkgo biloba* extract (GBE), and compared the results to those obtained by HPLC.

Flavonoids in Selected Botanicals

Ginkgo Biloba

Extracts of *Ginkgo biloba* leaves have been used since ancient times in China and since the 1960s in France and Germany to treat certain atherosclerotic diseases (*26*). *G. biloba* is one of the most highly characterized botanicals available today. It contains ginkgolides and bilobalides, terpene lactones (*26, 27*). Also present are benzoic acid derivatives, di-*trans*-poly-*cis*-octadecaprenol, ginkgolic acids, 2-hexenal, kynurenic acids (*N*-containing acids), proanthocyanidins, steroids, sterols, sugars, and waxes (*28*). Flavonol glycosides (Figures 1-3) are present, as are flavone glycosides (Figure 4), coumaric esters of kaempferol and quercetin (Figure 5), biflavones (Figure 6), and flavonol triglycosides (Figure 7) (*13, 18, 19, 22, 25, 26, 28-34*).

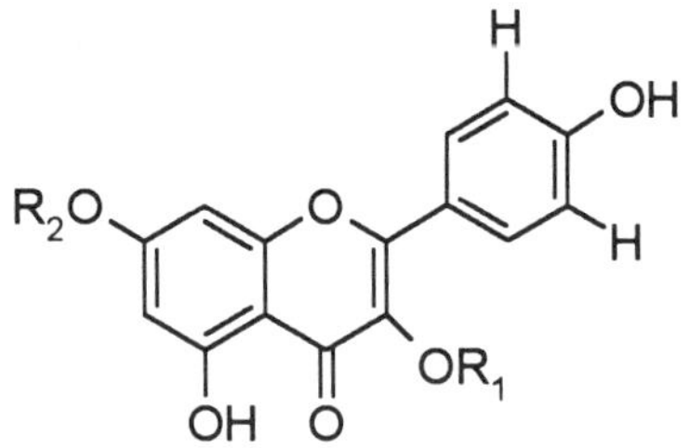

Compound	R1	R2
astragalin[a,b]	Glu	H
astragalin-6"-O-acetate[c]	Glu-6-O-acetate	H
kaempferol[d]	H	H
kaempferol-3-O-glucoside[b]	Glu	H
kaempferol-3-O-glucoside-7-O-rhamnoside[e]	Glu	Rham
kaempferol 3-O-(2"-glucosyl) rhamnoside[b]	Rham-Glu	H
kaempferol-3-O-rhamnoside[b]	Rham	H
kaempferol-3-O-rutinoside[b]	Rut	H
kaempferol-3,7-O-dirhamnoside[e]	Rham	Rham
kaempferol-7-O-glucoside[b]	H	Glu
tiliroside[e]	X''-p-coumaryl-glucose	H

[a] In *Equisetum arvense* (*13*). [b] In *Ginkgo biloba* (*13, 25, 28, 29, 31, 34*).
[c] In *Apocynum venetum* (*15*). [d] In *Centaurea erythraea* after hydrolysis (*50*).
[e] In *Tilia cordata* (*18*).

Figure 1. Kaempferol and kaempferol glycosides in several botanicals.

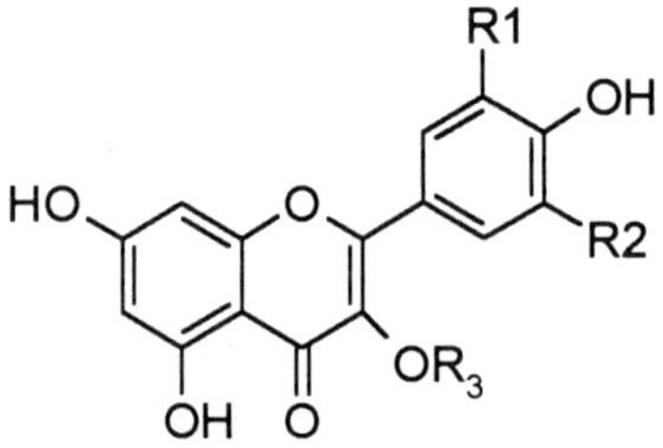

Compound	R1	R2	R3
3'-O-methylmyricetin 3-O-glucoside[a]	OMe	OH	Glu
3'-O-methylmyricetin 3-O-rutinoside[a]	OMe	OH	Rut
myricetin[a,b]	OH	OH	H
myricetin-3-O-rutinoside[a]	OH	OH	Rut
syringetin-3-O-rutinoside[a]	OMe	OMe	Rut

[a] In *Ginkgo biloba* (*28, 29, 34*). [b] In *Limonium sinense* (*51*).

Figure 2. Myricetin and myricetin glycosides in Ginkgo biloba and in Limonium sinense.

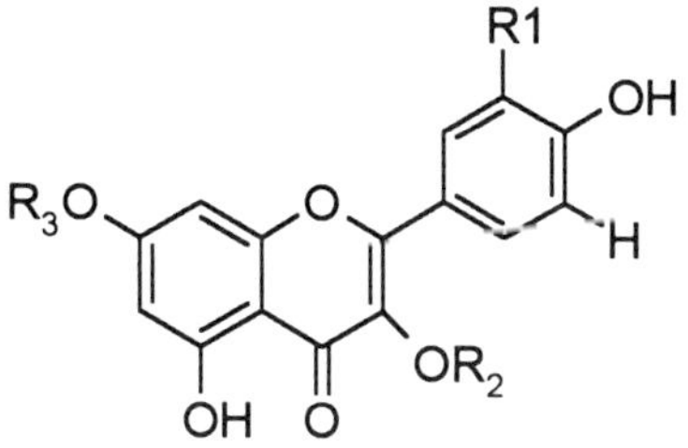

Compound	R1	R2	R3
avicularin[a]	OH	Ara	H
hyperoside[a-c]	OH	Gal	H
isoquercitrin[a-d]	OH	Glu	H
isoquercitrin-6"-O-acetate[b]	OH	Glu-6-O-acetate	H
isorhamnetin[e]	OMe	H	H
isorhamnetin-3-O-glucoside[a]	OMe	Glu	H
isorhamnetin-3-O-rutinoside[a]	OMe	Rut	H
isorhamnetin-3-O-2^G-rhamnosylrutinoside[e]	OMe	2^G-rhamnosyl-rutinose	H
miquelianin[c]	OH	glucuronose	H
quercetin[a-c,f-h]	OH	H	H
quercetin-3-O-glucoside[a]	OH	Glu	H
quercetin-3-O-glucoside-7-O-rhamnoside[e,i]	OH	Glu	Rham
quercetin-3-O-sophoroside[d]	OH	sophorose	H
quercetin-3-O-(2"-glucosyl)rhamnoside[a]	OH	Rham-Glu	H
quercetin-3,7-O-dirhamnoside[e,i]	OH	Rham	Rham
quercitrin[a,c]	OH	Rham	H
rutin[a,c]	OH	Rut	H

[a] In *Ginkgo biloba* (*13, 22, 25, 28, 29, 31, 34*). [b] In *Apocynum venetum* (*15*).
[c] In *Hypericum perforatum* (*38, 42*). [d] In *Equisetum arvense* (*13*).
[e] In *Calendula officinalis* (*18*). [f] In *Centaurea erythraea* after hydrolysis (*50*).
[g] In *Limonium sinense*. [h] In *Hypericum androsaemum* after hydrolysis (*50*).
[i] In *Tilia cordata* (*18*).

Figure 3. Quercetin and quercetin glycosides in several botanicals.

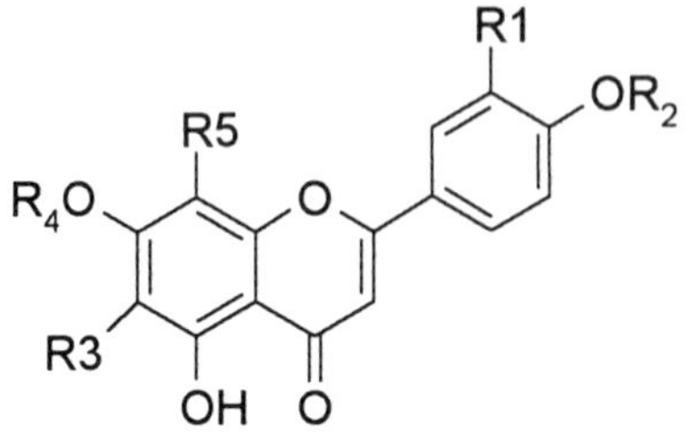

Compound	R1	R2	R3	R4	R5
acacetin[a]	H	Me	H	H	H
apigenin[b]	H	H	H	H	H
apigenin-7-O-β-acetylglucoside[c]	H	H	H	6"-O-acetyl-Glu	H
apigenin-7-O-β-glucoside[c-e]	H	H	H	Glu	H
apiin[d]	H	H	H	apiose-β-D-Glu	H
isovitexin[f]	H	H	β-D-Glu	H	H
isovitexin-2"-β-glucoside[e]	H	H	β-D-Glu-β-D-Glu	H	H
luteolin[b]	OH	H	H	H	H
luteolin-3'-O-glucoside[e]	OGlu	H	H	H	H
luteolin-7-O-glucoside[d]	OH	H	H	Glu	H
vitexin[f,g]	H	H	H	H	β-D-Glu
vitexin-rhamnoside[g]	H	H	H	H	Glu-Rham
vitexin-4'-O-rhamnoside[g]	H	Rham	H	H	β-D-Glu

[a] In *Ginkgo biloba* after hydrolysis (*44*). [b] In *Limonium sinense* (*51*).
[c] In *Matricaria chamomilla* (*49*). [d] In *Anthemis nobilis* (*13*).
[e] In *Ginkgo biloba* (*29*). [f] In *Passiflora incarnata* (*45, 46*).
[g] In *Crataegus monogyna* (*45, 46*).

Figure 4. Flavones and flavone glycosides in several botanicals.

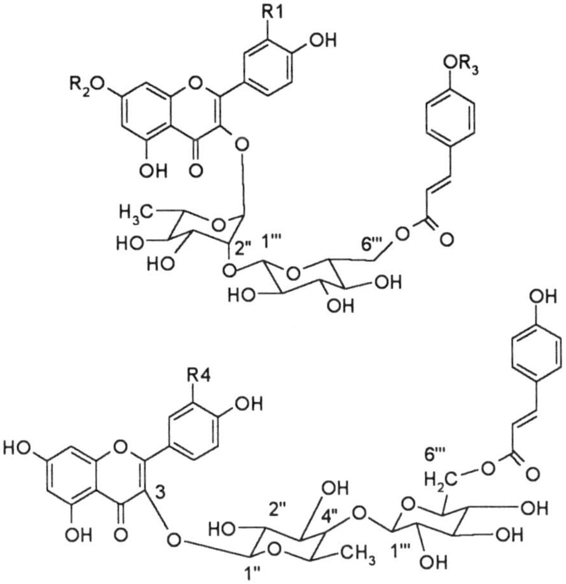

Compound	R1	R2	R3	R4
quercetin-3-*O*-[2"-*O*-[6'''-*O*-(*p*-hydroxy-trans-cinnamoyl)-*β*-D-glucosyl]-*α*-L-rhamnoside]	OH	H	H	---
kaempferol-3-*O*-[2"-*O*-[6'''-*O*-(*p*-hydroxy-trans-cinnamoyl)-*β*-D-glucosyl]- *α*-L-rhamnoside]	H	H	H	---
quercetin-3-*O*-(2"-*O*-[6'''-*O*-[*p*-(*β*-D-glucosyloxy)-trans-cinnamoyl]-*β*-D-glucosyl]-*α*-L-rhamnoside)	OH	H	Glu	---
kaempferol-3-*O*-(2"-*O*-[6'''-*O*-[*p*-(*β*-D-glucosyloxy)-trans-cinnamoyl]-*β*-D-glucosyl]-*α*-L-rhamnoside)	H	H	Glu	---
quercetin-3-*O*-[2"-*O*-[6'''-*O*-(*p*-hydroxy-trans-cinnamoyl-*β*-D-glucosyl]-*α*-L-rhamnosyl]-7-*O*-*β*-D-glucoside	OH	Glu	H	---
kaempferol-3-*O*-*α*-(6'''-*p*-coumaroylglucosyl-*β*-1,4-rhamnoside)	---	---	---	H
quercetin-3-*O*-α-(6'''-*p*-coumaroylglucosyl-*β*-1,4-rhamnoside)	---	---	---	OH

Figure 5.   Coumaric esters of flavonol glycosides in Ginkgo biloba (28, 31, 32).

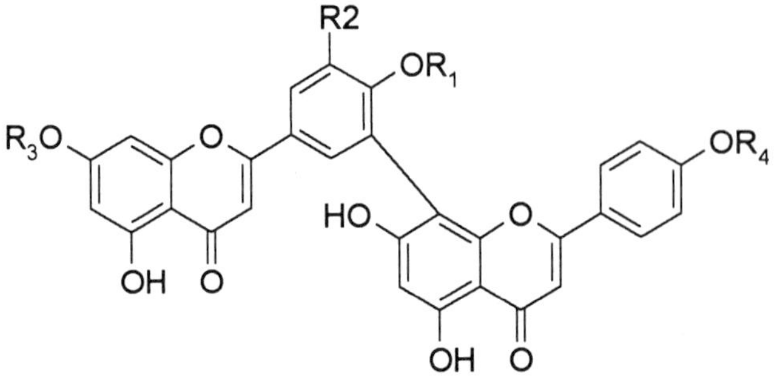

Compound[a]	R1	R2	R3	R4
amentoflavone	H	H	H	H
bilobetin	Me	H	H	H
ginkgetin	Me	H	Me	H
isoginkgetin	Me	H	H	Me
5'-methoxybilobetin	Me	OMe	H	H
sciadopitysin	Me	H	Me	Me
sequojaflavone	H	H	Me	H

[a] All compounds found in *Ginkgo biloba* (*28, 34*); amentoflavone found in *Hypericum perforatum* (*41*).

Figure 6. Structures of biflavones in Ginkgo biloba and Hypericum perforatum.

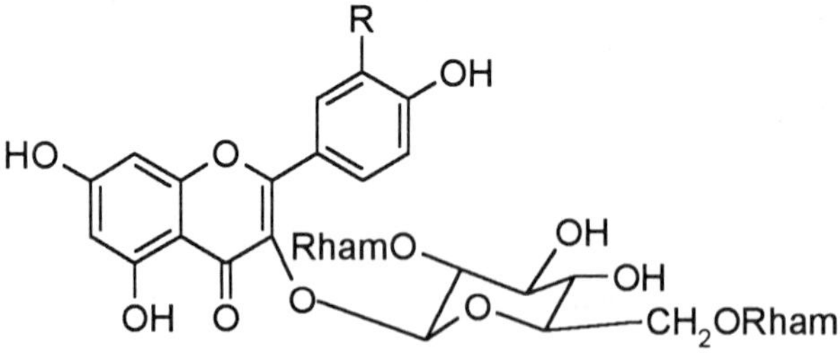

Compound	R
kaempferol-3-O-[2"-O, 6"-O-bis(α-L-rhamnosyl)-β-D-glucosyl]	H
quercetin-3-O-[2"-O, 6"-O-bis(α-L-rhamnosyl)-β-D-glucosyl]	OH
isorhamnetin-3-O-[2"-O, 6"-O-bis(α-L-rhamnosyl)-β-D-glucosyl]	OMe

Figure 7. Flavonol triglycosides in Ginkgo biloba (28).

Pycnogenol

Pycnogenol, used as a supplement because of radical scavenging abilities, is a mixture of catechins and procyanidins from the bark of *Pinus maritima,* the maritime pine tree (*17, 36*). Patented (Horphag Research, Geneva, Switzerland), it is prepared from trees in Landes de Gascogne in France, in Bay of Biscay, where the proper environmental influences on the trees appear unique.

Procyanidins inhibit synthesis of thromboxane A-2, a proaggregatory compound (*37*). In dose-response human studies, both aspirin and Pycnogenol showed activity against smoking-induced platelet aggregation. However, aspirin caused bleeding in some of the subjects. Pycnogenol reduced edema formation, and has other good effects, such as vaso-protection, and strengthening capillary resistance and walls.

Cossins and colleagues (*5*) tested various flavonoids including Pycnogenol, noticing that they can extend the lifetime of the ascorbate radical. It appears that flavonoids could regenerate ascorbate from the ascorbate radical. Since ascorbate regenerates Vitamin E, flavonoid sources such as Pycnogenol could help regenerate Vitamin E. The hydroxycinnamic acids caffeic, coumaric, and ferulic acids found in Pycnogenol also possess antioxidant activity.

St. John's Wort

Hypericum perforatum, found in Asia, northern Africa, and Europe (*38*), has been brought to the Americas, East Asia, New Zealand, and Australia (*39*). Formerly used for its healing and anti-inflammatory properties, it is now used as an anti-depressant (*38*). Also called St. John's wort and Klamath weed (*40*), it contains chlorogenic acid, the naphthodianthrones hypericin and pseudohypericin, and the phloroglucinols adhyperforin and hyperforin (*38*). Flavonoids include hyperoside, isoquercitrin, 3,3',4',5,7-pentahydroxyflavanone 7-*O*-rhamnopyranoside, quercetin, quercitrin, and rutin (Figure 1) (*38*); amentoflavone (Figure 6) (*41*); miquelianin (Figure 3) and the flavanonol (*9*) glycoside astilbin (Figure 8) (*42*); and 3,8"-biapigenin (Figure 9) (*43*). St. John's wort oil, *Hyperici oleum,* used to heal wounds, has no hypericin (*21*), but adhyperforin and hyperforin were still found. Quercetin and xanthones were found after SPE.

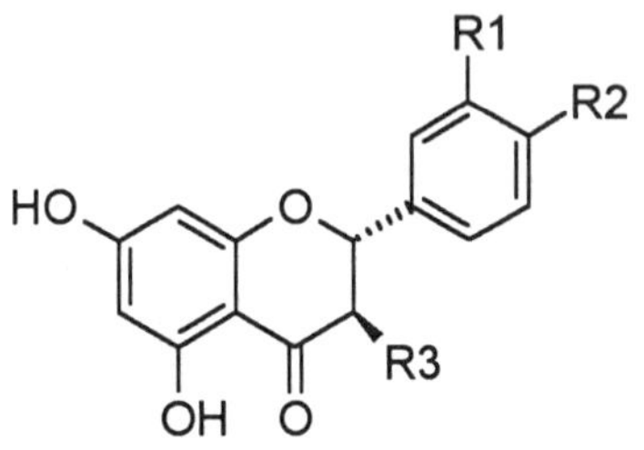

32

Compound	R1	R2	R3
astilbin	OH	OH	ORham
eriodictyol	OH	OH	H
hesperetin	OH	OMe	H
homoeriodictyol	OMe	OH	H
naringenin	H	OH	H

Figure 8.   The flavanonol (9) glycoside astilbin and four flavanones (structures from 6).

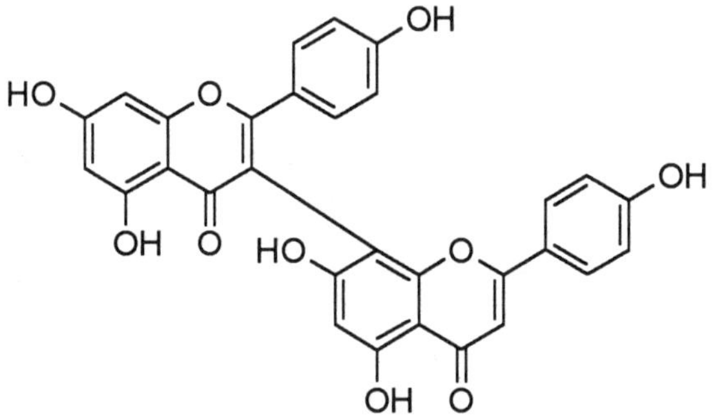

Figure 9. 3,8"-Biapigenin (38) found in St. John's Wort.

Other Botanicals

Apigenin and quercetin (Figures 3-4) were found after deglycosylation in *Achillea millefolium* (*44*), or yarrow. Deglycosylation of *Betula alba* (birch) yielded myricetin and quercetin (Figures 2-3). The flavone glycosides apigenin-7-*O*-glucoside, apiin (apigenin-7-*O*-apioglucoside), and luteolin-7-*O*-glucoside (Figure 4) were found in *Anthemis nobilis* (*13*), or Roman chamomile. Deglycosylation of *Calendula officinalis,* or garden marigold, yielded isorhamnetin and quercetin (Figure 3) (*44*). Isorhamnetin-3-*O*-2^G-rhamnosyl-rutinoside (Figure 3) was found under non-hydrolytic conditions (*18*).

Anti-hypertensive, sedative, anti-hyperlipemic, and diuretic properties are found in *Apocynum venetum* (dogbane leaf), whose leaves are used for making teas in Japan and northern China (*15*). Quercetin, four flavonol glycosides (Figures 1,3), and four catechins (Figure 10) were found in *A. venetum,* along with apocynin A, B, C, and D, catechin-[8,7-*e*]-4α-(3,4-dihydroxyphenyl)-dihydro-2(3*H*)-pyranone, and cinchonain Ia (Figures 10-13). The flavonol glycosides astragalin (Figure 1), isoquercitrin, and quercetin-3-*O*-sophoroside (Figure 3) were found in *Equisetum arvense* (*13*) (horsetail).

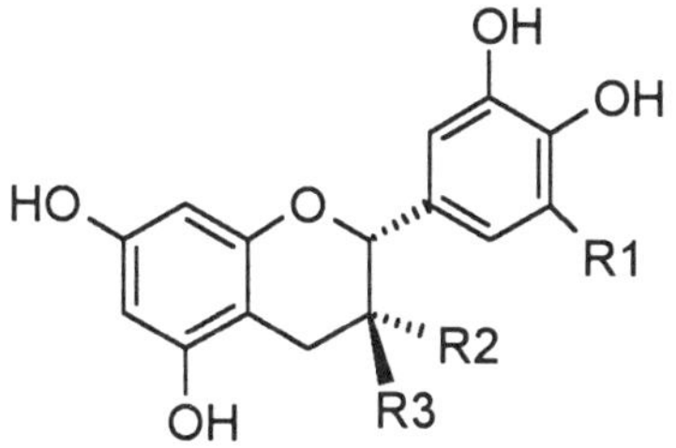

compound	R1	R2	R3
(+)-catechin	H	H	OH
(−)-epicatechin	H	OH	H
(−)-epigallocatechin	OH	OH	H
(+)-gallocatechin	OH	H	OH

Figure 10. Flavan-3-ols in Apocynum venetum (15).

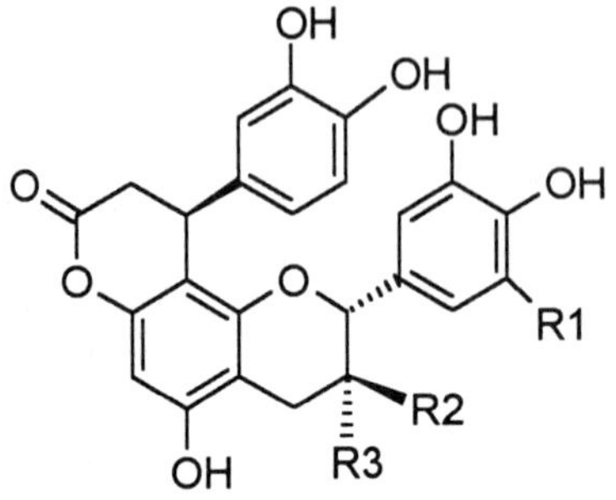

34

compound	R1	R2	R3
apocynin A	H	OH	H
apocynin C	OH	H	OH
cinchonain Ia	H	H	OH

Figure 11. Flavan-3-ols in Apocynum venetum (15).

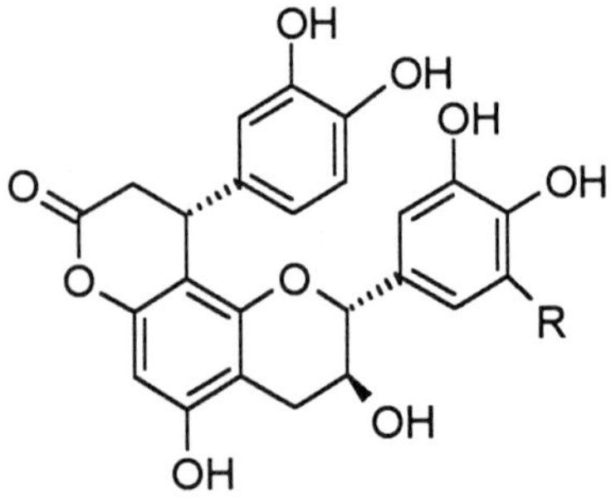

compound	R
apocynin B	OH
catechin-[8,7-e]-4α-(3,4-dihydroxy-phenyl)-dihydro-2(3H)-pyranone	H

Figure 12. Flavan-3-ols in Apocynum venetum (15).

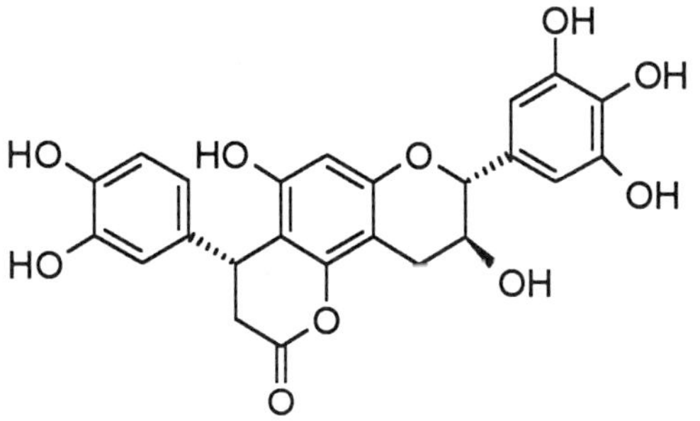

Figure 13. Apocynin D (15).

Procyanidins and flavonoids are found in *Crataegus* ssp. (hawthorn) (*14*). Dried flowering tops of *Crataegus monogyna* (singleseed hawthorn) and *C. laevigata* are used to produce a drug used for improvement of heart function. *C. monogyna* and *Passiflora incarnata* are used for sedative preparations, and contain *C*-glycosides of flavones (Figure 4) (*45, 46*).

Hippophae rhamnoides, or sea buckthorn, used as early as 900 AD in Tibet for medicinal applications and later domesticated (*47*), contains L-ascorbic acid. Kaempferol and quercetin (Figure 3) were detected after hydrolysis in aqueous methanolic HCl.

Ilex dumosa is a substitute or adulterant of *Ilex paraguariensis,* or yerba mate. *I. dumosa* was collected in Paraná, Brazil, its original habitat. *I. dumosa* contains caffeic acid, chlorogenic acid, cynarin, three isochlorogenic isomers, quercetin, and rutin (Figure 3) (*48*). Trace amounts of the xanthines caffeine and theophylline were detected, but no theobromine.

Apigenin-7-β-D-glucoside and apigenin-7-β-D-(6"-*O*-acetyl)glucoside (Figure 4), and the coumarins herniarin and umbelliferone (Figure 14), are the main components in the flavone and coumarin fractions in flowers of *Matricaria chamomilla* (chamomile) (*46, 49*). Flavones and coumarins impart spasmolytic and antiphlogistic properties to the flowers. Extracts of *M. chamomilla* have also been used in cosmetics.

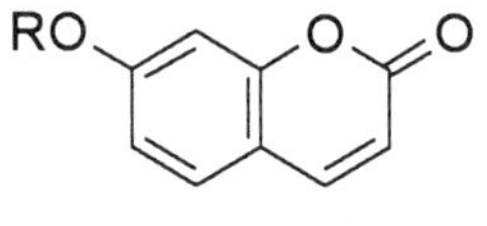

Figure 14. Herniarin (R = Me) and umbelliferone (R = H) (46, 49) found in Matricaria chamomilla.

36

Brazilians drink infusions of *Maytenus aquifolium* (espinheira-santa in Portuguese) to combat stomach diseases (*16*) The triterpenes friedelin and friedelan-3-ol have been isolated from its extracts, as have two flavonol tetraglycosides (Figure 15).

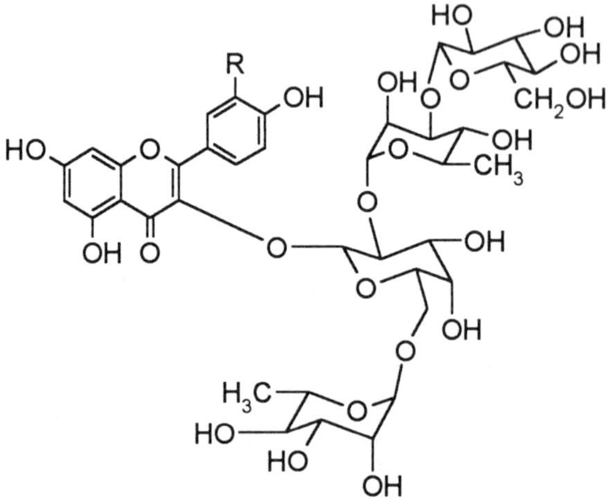

Compound	R
kaempferol 3-*O*-α-L -rhamnopyranosyl(1→6)-*O*-[β-D-glucopyranosyl (1→3)-*O*-α-L -rhamnopyranosyl(1→2)]-*O*-β-D-galactopyranoside	H
quercetin 3-*O*-α-L -rhamnopyranosyl(1→6)-*O*-[β-D-glucopyranosyl (1→3)-*O*-α-L -rhamnopyranosyl(1→2)]-*O*-β-D-galactopyranoside	OH

Figure 15. Flavonol tetraglycosides from Maytenus aquifolium (16).

Pietta and colleagues (*18*) found isoquercitrin, kaempferol-3,7-*O*-dirhamnoside, kaempferol-3-*O*-glucoside-7-*O*-rhamnoside, quercetin-3,7-*O*-dirhamnoside, quercetin-3-*O*-glucoside-7-rhamnoside, and tiliroside (Figures 1,3) in the leaves of *Tilia cordata*, or littleleaf linden.

Andrade and colleagues (*50*) chromatographed *Centaurea erythraea*, *Cynara cardunculus* (Cardoon aster), *Hypericum androsaemum* (Tutsan), *Lavandula officinalis* (lavender), *Lippia citriodora* (herb Louisa), *Mentha piperita* (peppermint), and *Salvia officinalis* (garden sage). The flavanones eriodictyol and hesperetin (Figure 8) were found in *M. piperita* after hydrolysis. The flavanones eriodictyol, homoeriodictyol, and naringenin (Figure 8) were among 20 flavonoids (Figures 2,3,4,16,17) identified by Lin and Chou (*51*) in *Limonium sinense*.

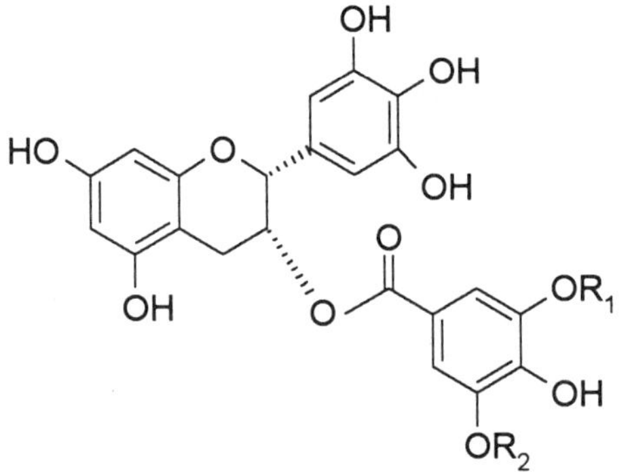

compound	R1	R2
(–)-epigallocatechin-3-*O*-gallate	H	H
(–)-epigallocatechin-3-*O*-(3'-*O*-methyl)-gallate	Me	H
(–)-epigallocatechin-3-*O*-(3',5'-di-*O*-methyl)-gallate	Me	Me

Figure 16. Flavan-3-ols in Limonium sinense (51).

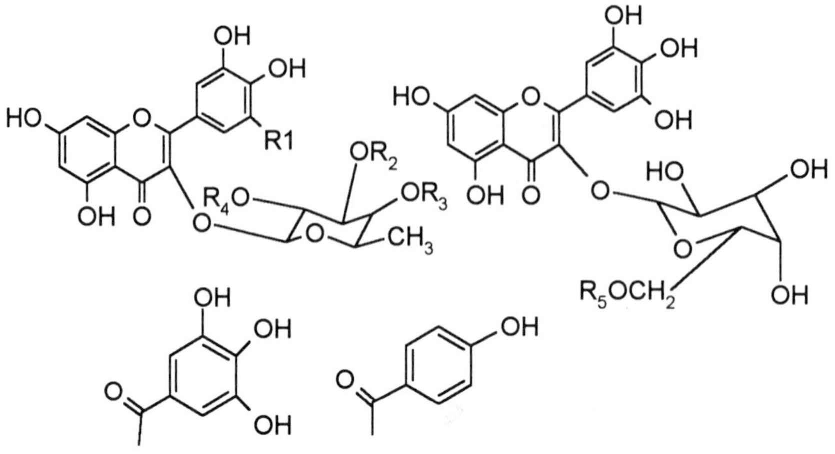

I

II

Compound	R1	R2	R3	R4	R5
myricetin-3-*O*-β-galactopyranoside	---	---	---	---	H
myricetin-3-*O*-(6"-*O*-galloyl)-β-galactopyranoside	---	---	---	---	I
myricetin-3-*O*-(2"-*O*-galloyl)-α-rhamnopyranoside	OH	H	H	I	---
myricetin-3-*O*-(3"-*O*-galloyl)-α-rhamnopyranoside	OH	I	H	H	---
myricetin-3-*O*-(4"-*O*-galloyl)-α-rhamnopyranoside	OH	H	I	H	---
myricetin-3-*O*-(2"-*O*-p-hydroxybenzoyl)-α-rhamnopyranoside	OH	H	H	II	---
myricetin-3-*O*-α-rhamnopyranoside	OH	H	H	H	---
quercetin-3-*O*-α-rhamnopyranoside	H	H	H	H	---
quercetin-3-*O*-(2"-*O*-galloyl)-α-rhamnopyranoside	H	H	H	I	---

Figure 17. Flavonol glycosides in Limonium sinense (51). Myricetin-3-O-β-arabinoside also found.

Isoflavones are also found in a limited number of botanicals. For example, *Sophora japonica,* the Chinese scholar tree or Japanese pagoda tree, contains the isoflavone genistein (Figure 18), the flavonoids quercetin and rutin (Figure 3), and sophorabioside and sophoricioside (52). *Ononis spinosa*, or spiny rest harrow, contains the isoflavones biochanin A, formononetin, and genistein (Figure 18) and the flavonols kaempferol and rutin (Figures 1,3) (53).

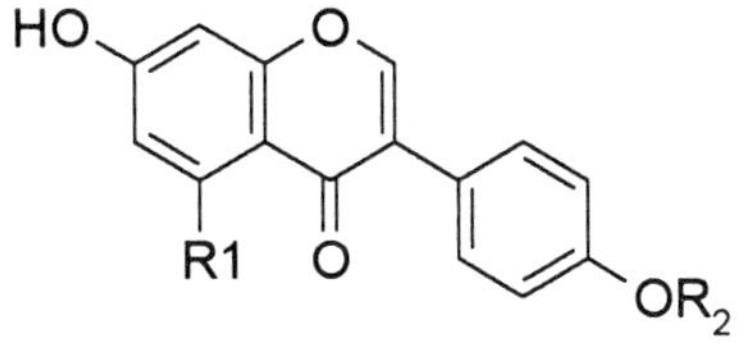

compound	R1	R2
biochanin A	OH	Me
formononetin	H	Me
genistein	OH	H

Figure 18. Isoflavones (structures from 6) in Ononis spinosa (53) and Sophora japonica (genistein only) (52).

Conclusion

Botanical preparations are generally taken from the leaves or flowers. Botanical flavonoids are usually glycosides of flavones or flavonols, with some catechins and proanthocyanidins found in Pycnogenol, which comes from tree bark. Catechins and flavanone glycosides are less common.

HPLC is the method of choice for identification and quantitation of flavonoid glycosides and, after hydrolysis, aglycones. Other methods such as MECC have been successfully used, but are not as prevalent in the literature.

References

1. Hertog, M.G.L.; Hollman, P.C.H.; Katan, M.B.; Kromhout, D. *Nutr. Cancer* **1993,** *20,* 21-29.
2. Middleton, Jr., E. *Int. J. Pharmacognosy* **1996,** *34,* 344-348.
3. Rice-Evans, C.A.; Miller, N.J.; Paganga, G. *Free Radical Biol. Med.* **1996,** *20, 933-956.*
4. Brown, J.E.; Khodr, H.; Hider, H.; Rice-Evans, C.A. *Biochem. J.* **1998,** *330,* 1173-1178.

40

5. Cossins, E.; Lee, R.; Packer, L. *Biochem. Mol. Biol. Int.* **1998,***45*, 583-597.
6. Mabry, T.J.; Markham, K.R.; Thomas, M.B. *The Systematic Identification of Flavonoids.* Springer-Verlag: New York, 1970.
7. Daigle, D.J.; Conkerton, E.J. *J. Liq. Chromatogr.* **1983,** *6*, 105-118.
8. Daigle, D.J.; Conkerton, E.J. *J. Liq. Chromatogr.* **1988,** *11*, 309-325.
9. Robards, K.; Antolovich, M. *Analyst* **1997**, *122*, 11R-34R.
10. Merken, H.M.; Beecher, G.R. *J. Agric. Food Chem.* **2000**, *48*, 577-599.
11. Rice-Evans, C.A.; Packer, L., Eds. *Flavonoids in Health and Disease.* Marcel Dekker, Inc.: New York, 1998.
12. Pietta, P. In *Flavonoids in Health and Disease*; Rice-Evans, C.A., Packer, L., Eds.; Marcel Dekker, Inc.: New York, 1998; p 61-110.
13. Pietta, P.; Mauri, P.; Bruno, A.; Rava, A.; Manera, E.; Ceva, P. *J. Chromatogr.* **1991**, *553*, 223-231.
14. Rohr, G.E.; Meier, B.; Sticher, O. *Phytochem. Anal.* **2000**, *11*, 106-112.
15. Xiong, Q.; Fan, W.; Tezuka, Y.; Adnyana, I.K.; Stampoulis, P.; Hattori, M.; Namba, T.; Kadota, S. *Planta Med.* **2000**, *66*, 127-133.
16. Vilegas, W.; Sanommiya, M.; Rastrelli, L.; Pizza, C. *J. Agric. Food Chem.* **1999,** *47*, 403-406.
17. Packer, L.; Rimbach, G.; Virgili, F. *Free Radical Biol. Med.* **1999**, *27*, 704-724.
18. Pietta, P.; Facino, R.M; Carini, M.; Mauri, P. *J. Chromatogr., A* **1994,** *661*, 121-126.
19. Hahn-Deinstrop, E.; Koch, A. *Phytopharmaka BIOforum* **1998,** *7-8*, 428-434.
20. Jamshidi, A.; Adjvadi, Ml; Husain, S.W. *J. Planar Chromatogr.* **2000**, *13*, 57-59.
21. Maisenbacher, P.; Kovar, K.-A. *Planta Med.* **1992,** *58*, 351-354.
22. Vanhaelen, M.; Vanhaelen-Fastre, R. *J. Liq. Chromatogr.* **1988**, *11*, 2969-2975.
23. Yang, F.; Quan, J.; Zhang, T.Y.; Ito, Y. *J. Chromatogr., A* **1998,** *803*, 298-301.
24. Skoog, D.A.; Holler, F.J.; Nieman, T.A. *Principles of Instrumental Analysis,* Fifth Ed.; Harcourt Brace & Company: USA, 1998; 795.
25. Pietta, P.G.; Mauri, P.L.; Rava, A.; Sabbatini, G. *J. Chromatogr.* **1991**, *549*, 367-373.
26. Shi, H.; Niki, E. *Lipids* **1998**, *33*, 365-370.
27. Hasler, A.; Sticher, O.; Meier, B. *J. Chromatogr.* **1992,** *605*, 41-48.
28. Beek, T.A. van; Bombardelli, E.; Morazzoni, P.; Peterlongo, F. *Fitoterapia* **1998,** *69*, 195-244.
29. Chengzhang, W.; Xiang, C.; Weihong, T.; Qing, Y.; Zhaobang, S. *Chem. Ind. Forest Prod.* **1998,** *18*, 83-88.

30. Chi, J.D.; He, X.F.; Liu, A.R.; Xu, L.X. *Acta Pharmaceutica Sinica* **1997**, *32*, 625-628.
31. Hasler, A.; Gross, G.-A.; Meier, B.; Sticher, W. *Phytochemistry* **1992**, *31*, 1391-1394.
32. Nasr, C.; Haag-Berrurier, M.; Lobstein-Guth, A.; Anton, R. *Phytochemistry* **1986**, *25*, 770-771.
33. Nasr, C.; Lobstein-Guth, A.; Haag-Berrurier, M.; Anton, R. *Phytochemistry* **1987**, *26*, 2869-2870.
34. Packer, L.; Saliou, C.; Droy-Lefaix, M.-T.; Christen, Y. In *Flavonoids in Health and Disease*; Rice-Evans, C.A., Packer, L., Eds.; Marcel Dekker, Inc.: New York, 1998; p 303-341.
35. Akiba, S.; Kawauchi, T.; Oka, T.; Hashizume, T.; Sato, T. *Biochem. Mol. Biol. Int.* **1998**, *46*, 1243-1248.
36. Huynh, H.T.; Teel, R.W. *Cancer Lett.* **1998**, *132*, 135-139.
37. Pütter, M.; Grotemeyer, K.H.M.; Würthwein, G.; Araghi-Niknam, M.; Watson, R.R.; Hosseini, S.; Rohdewald, P. *Thromb. Res.* **1999**, *95*, 155-161.
38. Brolis, M.; Gabetta, B.; Fuzzati, N.; Pace, R.; Panzeri, F.; Peterlongo, F. *J. Chromatogr., A* **1998**, *825*, 9-16.
39. Franke, R.; Schenk, R.; Bauermann, U. *Proc. WOCMAP-2, Acta Hort.* **1999**, *502*, 167-173.
40. Upton, R. *HerbalGram.* **1997**, *40*, *32*, 3-38.
41. Kartnig, T.; Göbel, I.; Heydel, B. *Planta Med.* **1996**, *62*, 51-53.
42. Butterweck, V.; Jürgenliemk, G.; Nahrstedt, A.; Winterhoff, H. *Planta Med.* **2000**, *66*, 3-6.
43. Bombardelli, E.; Morazzoni, P. *Fitoterapia* **1995**, *66*, 43-68.
44. Menghinello, P.; Cucchiarini, L.; Palma, F.; Agostini, D.; Dachà, M.; Stocchi, V. *J. Liq. Chrom. & Rel. Technol.* **1999**, *22*, 3007-3018.
45. Pietta, P.; Manera, E.; Ceva, P. *J. Chromatogr.* **1986**, *357*, 233-238.
46. Pietta, P.; Mauri, P.; Manera, E.; Ceva, P.; Rava, A. *Chromatographia* **1989**, *27*, 509-512.
47. Jeppsson, N.; Gao, X. *Agric. Food Sci. Finland* **2000**, *9*, 17-22.
48. Filip, R.; López, P.G.; Ferraro, G.E. *Proc. WOCMAP-2, Acta Hort.* **1999**, *502*, 405-408.
49. Pietta, P.; Manera, E.; Ceva, P. *J. Chromatogr., A* **1987**, *404*, 279-281.
50. Andrade, P.B.; Seabra, R.M.; Valentão, P.; Areias, F. *J. Liq. Chromatogr. & Rel. Technol.*, **1998**, *21*, 2813-2820.
51. Lin, L.-C.; Chou, C.-J. *Planta Med.* **2000**, *66*, 382-383.
52. Liu, I.M.; Sheu, S.J. *Am. J. Chinese Med.* **1989**, *17*, 179-187.
53. Pietta, P.; Calatroni, A.; Cesare, Z. *J. Chromatogr.* **1983**, *280*, 172-175.

Chapter 4

Analysis of Anthocyanins in Nutraceuticals

Ronald E. Wrolstad[1], Robert W. Durst[1], M. Monica Giusti[2], and
Luis E. Rodriguez-Saona[3]

[1]Department of Food Science and Technology, Oregon State University,
Corvallis, OR 97331–6602
[2]Department of Nutrition and Food Science, University of Maryland,
College Park, MD 20742
[3]Joint Institute for Food Safety and Applied Nutrition, University
of Maryland–FDA, Washington DC 20204

Systematic methods for identification of anthocyanin pigments
and their quantitative measurement are well-established
because of their importance in food color quality and their
usefulness in chemotaxonomic investigations. Interest in their
accurate analysis has heightened because of their importance
in nutraceuticals and functional foods. Efficient extraction can
be achieved with cryogenic milling and acetone extraction.
Total anthocyanin content is easily measured with the pH
differential spectrophotometric method while HPLC with
external standard quantitation is an attractive alternative.
Additional spectral procedures provide indices for polymeric
color and browning. Reverse phase HPLC with uv-visible
diode array detection is the method of choice for tracking
qualitative changes. Methods for identification include acid
hydrolysis and saponification reactions as well as NMR and
electro-spray mass spectroscopy (ESMS).

Our laboratory has been analyzing anthocyanin pigments for years because of their importance to the color quality of fresh and processed foods, their usefulness in fruit juice authenticity investigations (*1*), and their application for use as natural colorants (*2*). Today, however, there is increased interest in their accurate analysis because of their importance as natural dietary antioxidants that effectively trap free radicals. They may serve beneficial roles in combating coronary heart disease (*3*), cancer (*4*), and diabetes (*5*) as well as help to improve visual acuity (*6*). It is generally accepted that there is an association between these diseases and the damaging effect of free radicals (*7*). Several investigators have reported a high correlation between antioxidant activity of various fruit extracts as measured by oxygen radical absorbing capacity (ORAC) and anthocyanin pigment content (*8,9,10*). With this interest in anthocyanin pigment content comes the need for their accurate qualitative and quantitative measurements in fruits and vegetables, their processed products, extracts, and nutraceutical preparations. Raw materials may be purchased on the basis of their pigment content, and purchase specifications and label declarations for nutraceuticals may call for measurement of total anthocyanins. While effective and accurate methods are available, there is a need for standardization of methodology and nomenclature in both the scientific and industrial sectors.

Anthocyanin Pigment Structure

Anthocyanins are water-soluble pigments responsible for the red to purple to blue colors of many fruits, vegetables, and flowers, and are also sometimes present in stems, roots and leaves. Over 300 have been identified in nature (*11*). A generalized structure of the anthocyanins appears in Figure 1. Structural variation in the B-ring with H, OH or OCH_3 substitiuton gives six different anthocyanidins. They vary with respect to λ_{max} and expressed color as well as chemical reactivity. The anthocyanidins as such do not occur in nature, rather they occur as glycosides. The anthocyanidins which are generated by acid hydrolysis of the glycosidic pigments are very unstable. Wang *et al.* (*8*) measured the ORAC of five of the anthocyanidins and reported the following decreasing order: cyanidin > malvidin > delphinidin > peonidin > pelargonidin. Glycosidic substitution accounts for much of the structural variation in nature with possibilities of substitution with mono-, di-, and tri-saccharides at the 3 and /or 5 positions. Glycosidic substitution at the 7 position occurs in some rare instances. Acylation of sugar substituents with aromatic or aliphatic organic acids is an additional possibility which adds complexity to anthocyanin structure. It is well established that acylation, particularly with cinnamic acids, will give much greater stability to the pigment (*12*) . Acylation with aliphatic acids often goes undetected since these esters are quite labile to acid hydrolysis which can occur during extraction and isolation.

Anthocyanins undergo reversible structural transformations with change in pH which is accompanied by dramatic changes in color (Figure 2.)

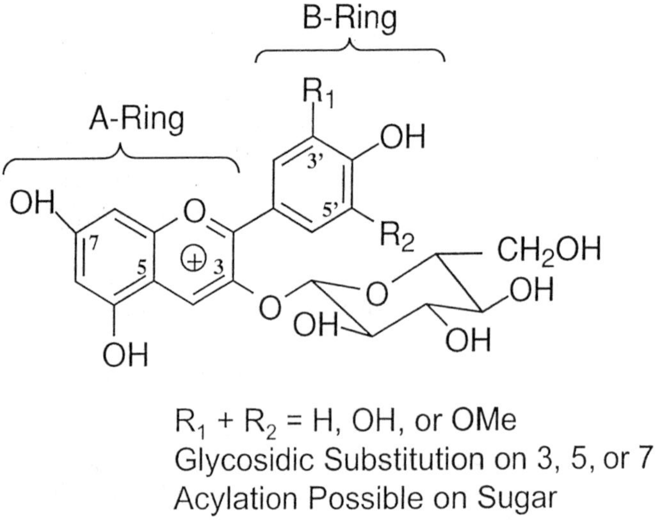

Figure 1. Generalized structure for anthocyanin pigments. Reproduced with permission from reference 2. Copyright 2000 Marcel Dekker, Inc.

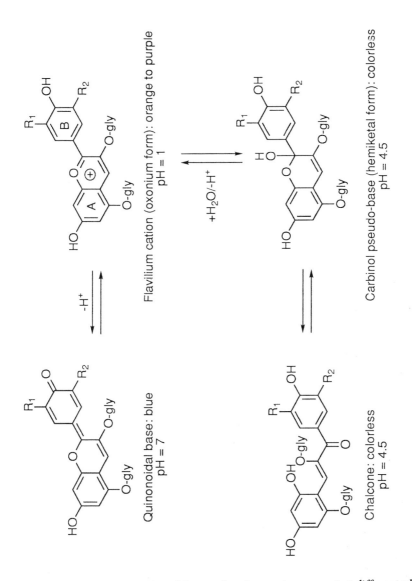

Figure 2. Predominant structural forms of anthocyanins present at different pH values. Reproduced with permission from reference 17. Copyright 2001 John Wiley & Sons.

These forms vary in stability with the flavylium cation being much more stable than the hemiketal and quinoidal base forms.

Extraction of Anthocyanins

Since anthocyanins are reactive compounds one needs to select an extraction method which maximizes pigment recovery, minimizes degradation and does not alter the natural (*in vivo*) state. Nevertheless, one can never be completely sure that the extracted pigment is exactly the one occurring *in vivo*. In a recent publication (*13*), we describe in detail procedures for extraction, isolation, and purification of anthocyanins. For most plant materials it is advantageous to cryogenically mill by freezing the sample in liquid nitrogen and powdering with a Waring Blendor or mill suitable for use under extremely low temperatures. Anthocyanin degradation is minimized by lowering the temperature and providing a nitrogen environment. The uniformity of the powdered sample facilitates handling and weighing and the high surface area of the small particles helps to maximize pigment recovery.

Methanol is the most common solvent used for anthocyanin extraction. The anthocyanins are highly soluble in methanol and polysaccharides and proteins are precipitated allowing for easy filtration. The low boiling point of methanol permits its removal with a rotary evaporator at a relatively low temperatures, e.g., 40 °C. It is recommended that the solvent be acidified since the oxonium form is much more stable and the positive charge increases polarity. Methanol containing 0.01% HCl is widely used but some workers favor the use of weaker acids such as formic, acetic or citric to prevent acid hydrolysis of pigment acylations with aliphatic acids (*11,13*). Use of trifluoroacetic acid (3%) is also advised since its volatility permits its removal during concentration (*11,13*).

Our preference is to use acetone as the solvent for extraction followed by partition with chloroform to give an aqueous phase (containing the anthocyanins, phenolics, sugars, organic acids) and an organic phase containing the bulk of the solvents and lipid materials. We have applied this procedure to a wide range of anthocyanin-containing plant materials and obtained high recoveries with minimal pigment degradation (*13,14,15*). The volume of the aqueous phase is essentially the quantity of water in the sample; thus dried materials will appropriately be extracted with 30% aqueous acetone. There is some residual acetone in the aqueous extract which can readily be removed with a rotary evaporator. The absence of a concentration step minimizes the risk of acid-dependent pigment degradation. The aqueous extract is free of lipids, carotenoids and chlorophyll pigments. It can be advantageous to have the anthocyanins in aqueous solution for further purification by solid-phase extraction with C-18 resins (*13,16*), spectrophotometric quantitation by the pH differential method (*17*) and injection for HPLC separation (*18*). We have compared the effectiveness of the acetone/chloroform method with the acidified methanol extraction procedure for twenty nutraceutical and natural colorant preparations. The materials were juices, extracts and dried powders derived from cranberries (*Vaccinium macrocarpon*, elderberries (*Sambucus nigra* L.), chokeberries (*Aronia melanocarpa*) and bilberries (*V. myrtillus* L.). The results are shown in Table I. Anthocyanin recoveries were in all cases, except for sample 19, higher with the use of acetone. Recoveries were typically twice as high with acetone extraction. We suspect that the higher recoveries may be due to reduced thermal degradation during rotary evaporator concentration as well as extraction efficiency of aqueous acetone.

Measurement of Total Anthocyanin

Total Anthocyanin Pigment by the pH Differential Method

Accurate measurement of total anthocyanin pigment content is probably the most critical measurement for anthocyanin-derived nutraceuticals. While there is no official standardized method, we recommend the spectrophotometric pH differential method (*17,19*). Many plant scientists will determine total anthocyanin pigment by measuring the absorptivity of an acidified methanol or ethanol extract at a single wavelength. This is possible since anthocyanins have a characteristic absorption band in the 490-550 nm region of the visible spectra which is distinct from the absorption bands of other phenolics and pigments in the crude extract. This simpler procedure is inappropriate, however, for most processed foods and nutraceutical preparations because of possible interference from anthocyanin degradation products and melanoidin pigments which are produced from non-enzymatic and enzymatic browning reactions (*20*). The pH differential method takes advantage of the reversible structural transformations which anthocyanins undergo with change in pH (Figure 2). Essentially 100% of the pigment exists in the colored oxonium form at pH 1 while the colorless hemi-ketal form predominates at pH 4.5. By dissolving the sample in pH 1 and 4.5 buffers and measuring the absorbance at the wavelength of maximum absorption the total anthocyanin pigment content is measured by the following formula:

Monomeric anthocyanin pigment (mg/liter) = (A x MW x DF x 1000) / (ε x *l*)

A = difference in absorbance between the pH 1.0 and 4.5 solutions
MW = molecular weight
DF = dilution factor

Table I. Anthocyanin (ACN) Pigment Content (mg/100g) of 20
Nutraceutical/Natural Colorant Preparations[a] Comparing Acetone
vs Methanol Extraction, and pH Differential vs HPLC.

Sample	Total ACN, pH dif. MeOH Extraction	Total ACN, pH diff. Acetone Extraction	Total ACN, HPLC Acetone Extraction
1		86.2	78.3
2		60.4	52.1
3		62.6	40.6
4		669	604
5	4701	6,560	5,931
6	315	760	589
7	10.5	23.7	23.7
8	31.4	273	260
9	477	1,388	1,586
10	22.3	49.3	35.1
11	64.5	200	222
12	87.7	175	127
13	53.6	134	103
14	3,881	5,496	4,765
15	11,305	11,945	11,556
16	7,595	8,720	8,100
17	916	1,678	2,098
18	9,840	10,608	9,014
19	326	228	254
20	442	935	772

[a]20 anthocyanin-based nutraceutical preparations derived from elderberries (*Sambucus nigra* L.), chokeberries, (*Aronia melanocarpa*), cranberries (*Vaccinium macrocarpon*) and bilberries (*Vaccinium myrtillus* L.). Samples 1-4 were juice concentrates, the remainder were dried preparations.

ε = molar absorptivity
l = pathlength

The term "monomeric" is included to emphasize that the measurement will not incorporate the degraded polymeric anthocyanins which do not form the colorless hemi-ketal structure and remain colored at pH 4.5.

The calculation requires selection of an appropriate molar extinction coefficient. Table II lists molar absorptivity values for some selected anthocyanins as reported in the literature. [A more extensive table is given in the recent publication of Giusti and Wrolstad (17)]. The absorptivities for many anthocyanins have not been determined and there is lack of uniformity for the values that have been reported for the same pigments by different workers. This is not surprising because of the difficulty in preparing crystalline anthocyanins, free from impurities, in sufficient quantities to allow reliable weighing under optimal conditions (21). The extinction coefficient is not only dependent on the structure, but also on the solvent used. Many have been determined in methanol or ethanol containing HCl and others have been determined in aqueous systems such as 0.1 N HCl. Since the coefficient selected should have been determined in the same solvent as that used for the spectral measurement, those values reported for aqueous systems are most appropriate for the above pH differential procedure. It is custom to have anthocyanin calculations based on the major pigment present in the sample. This may not be possible since an extinction coefficient for that pigment may have not yet been determined. One recourse is to calculate pigment content as cyanidin-3-glucoside as it is the most abundant anthocyanin in nature. The molecular weights of anthocyanins vary tremendously, e.g., from 433.2 for pelargonidin-3-glucoside (major pigment in strawberries) to 1019.2 for pelargonidin-3-sophoroside-5-glucoside acylated with ferulic and malonic acids (major pigment in radishes). Thus the molecular weight used in the calculation will have an obvious effect on the resulting value. There is some justification for calculating anthocyanin content of nutraceutical preparations as cyanidin-3-glucoside, or perhaps malvidin-3-glucoside, since this will facilitate comparison of anthocyanin content among different sources and preparations. At any rate, it is paramount that one reports the molar absorptivity and molecular weight used in the determination since this will permit other workers to re-calculate and normalize pigment data for comparative purposes.

Determination of Total Anthocyanins by HPLC

HPLC is routinely used for separation of anthocyanins and the availability of pure anthocyanin standards makes quantitation by the external standard method possible. Extrasynthese (Genay, France), Polyphenols As (Sandnes, Norway), and Funakoshi Co. (Tokyo, Japan) are commercial firms

Table II. Reported Molar Absorptivity Values for Anthocyanins

Anthocyanin	Solvent	λmax (nm)	Molar Absorbtivity (ε)	Ref
Cyanidin (cyd)	0.1% HCl in ethanol	547	34,700	22
Cyd-3-glu	Aqueous buffer, pH 1	510	26,900	23
Cyd-3-glu	0.1 N HCl	520	25,740	24
Cyd-3-glu	1% HCl in methanol	530	34,300	25
Cyd-3-gal	0.1% HCl in methanol	530	34,300	25
Cyd-3-gal	15:85 0.1 N HCl/ethanol	535	46,200	26
Cyd-3-gal	HCl in methanol	530	30,200	27
Malvidin (Mvd)	0.1% HCl in methanol	520	37,200	28
Mvd	0.1% HCl in ethanol	557	36,200	22
Mvd-3-glu	0.1 N HCl	536	28,000	29
Mvd-3-glu	0.1% HCl in methanol	546	17,800	30
Mvd-3-glu	0.1% HCl in methanol	538	29,500	31
Mvd-3-5-diglu	0.1 N HCl	520	37,700	29
Mvd-3-5-diglu	0.1% HCl in ethanol	545	10,300	22
Pelargonidin (Pgd)	Aqueous buffer, pH 1.0	505	18,420	32
Pgd	0.1% HCl in methanol	524	19,780	32
Pgd	0.1% HCl in ethanol	505	17,800	28
Pgd-3-glu	1% HCl in water	496	27,300	33
Pgd-3-glu	1% HCl	513	22,390	27
Pgd-3-glu	1% HCl in ethanol	516	31,620	27

which market a number of anthocyanin pigment standards. In our laboratory, we measured the total anthocyanin content of the same 20 nutraceutical and natural colorant prepartions previously described by separating the pigments by HPLC, measuring the individual and total peak areas at 520 nm, and quantitating individual and total anthocyanins using commercial cyanidin-3-glucoside as an external standard We also measured the total monomeric anthocyanin pigment content of these cranberry, elderberry, chokeberry and bilberry-derived samples by the pH differential method. The comparative results which are shown in Table I are of the same order. Values are typically slightly higher for the pH differential method, but the close agreement gives validity to both of the analytical methods.

Indices for Polymeric Color and Browning

Much of the color of wines, fruit juices and other anthocyanin-pigmented processed foods is due to polymeric pigments where the anthocyanins have condensed with other flavonoids, the site for cross-linking occurring at the 4 position of the anthocyanin. A new class of anthocyanin pigments has recently been identified in wines where the anthocyanins are substituted at the 4 position with vinyl or $C_3H_2O_2$ (34). Melanoidin pigments from enzymatic and non-enzymatic browning will also contribute to the color of these foods. Monomeric anthocyanin pigments will combine with bisulfite to form a colorless adduct, the sulfonic acid substition occurring at the 4 position (Figure 3). Polymeric anthocyanins and the Vitisin A-type wine pigments will not undergo this bleaching reaction since the 4 position is blocked. Also, melanoidin pigments are not substantially bleached by SO_2. Thus reaction with bisulfite provides a simple method for getting indices for polymeric color and browning (17,19). Aqueous samples at an appropriate dilution are treated respectively with metabisulsfite reagent and water. Absorbance measurements are made at $\lambda_{vis-max}$ and 420 nm of the bisulfite-treated and control samples. Color density is the sum of the absorbances at the $\lambda_{vis-max}$ and at 420 nm of the control sample, while polymeric color is the same measurement for the bisulfite treated sample. A measure of percent polymeric color is obtained as the ratio between these two indexes. The absorbance at 420 nm of the bisulfite-treated sample is an index for browning.

HPLC Separation of Anthocyanins

Reverse-phase HPLC of Anthocyanin Pigment

HPLC has become the method of choice for analyzing anthocyanin pigments. It offers several advantages over paper or thin-layer chromatography

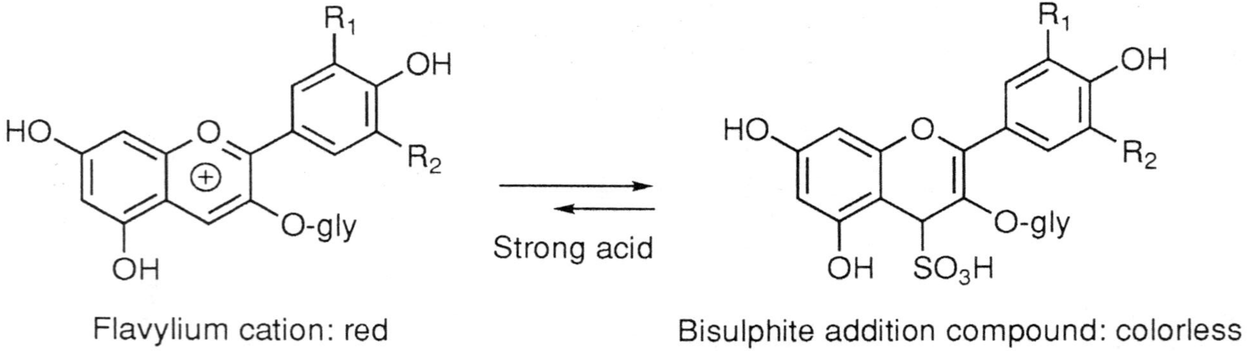
R1
OH
HO
O
R2
O-gly
OH
Flavylium cation: red
Strong acid
R1
OH
HO
O
R2
O-gly
OH
SO3H
Bisulphite addition compound: colorless

Figure 3. Formation of colorless anthocyanin-sulfonic acid adducts. Reproduced with permission from reference 17. Copyright 2001 John Wiley & Sons.

including greater resolution, shorter analysis times and easier quantitation. Another advantage of reverse-phase chromatography is the predictability of elution order based on polarity. Triglycosides typically elute before diglycosides which elute before monoglycosides. An exception to this generalization are the rutinosides which have longer retention times than most monoglycosides because of the non-polarity imparted by the C-6 CH_3 group of rhamnose. Hexose glycosides elute before pentose glycosides. The elution order of the aglycons can be predicted from the number of hydrophilic phenolic and hydrophobic methoxyl groups, the order being delphinidin < cyanidin < petunidin < pelargonidin < peonidin < malvidin. Acylation of sugar substituents with aliphatic and aromatic organic acids substantially reduces pigment polarity with much longer retention times resulting.

Two different types of HPLC columns are useful for anthocyanin separation, Silica C_{18} and Polymeric C_{18} columns. The former have shorter analysis times and are appropriate for the simpler anthocyanin glycosides while the Polymeric-based columns are more effective for more complex, non-polar acylated anthocyanin mixtures. Figure 4 shows the separation of cranberry anthocyanins on a Silica C_{18} column in less than 20 minutes while resolution of the Concord grape juice anthocyanins on a Polymeric C_{18} column (Figure 5) requires more than an hour. It is critical that the mobile phase be at low pH to ensure that the anthocyanins are in the oxonium form. Gradient systems are used with both types of columns. The initial composition of the mobile phase is typically phosphate buffer or acidified water with decreasing polarity in the gradient from increasing amounts of acetonitrile or methanol. The following references can be consulted for specific chromatographic conditions: *18,11,14, 35,36.*

Sample Preparation

For most applications, dilution of the sample with water, dilute HCl or phosphoric acid and filtering through a 0.45 μm filter is all that is required. Injection of a methanolic solution of anthocyanins will have relatively little effect on chromatographic behavior since the small injection volume (typically 50 μL) will have little effect on the composition of the mobile phase. In cases where the sample contains considerable degraded or polymeric pigment because of processing or storage abuse, sample clean-up is recommended where the anthocyanins are isolated by solid-phase extraction using C_{18} cartridges (*14,16, 18,37*). It is advisable to use this procedure for complex anthocyanin matrices since better chromatographic resolution will be achieved. This is particularly the case for separations using the Polymeric C_{18} columns.

Sample Detection

Since anthocyanins absorb in the visible spectrum, a variable wavelength detector is sufficient for sample detection. A uv-visible diode array detector,

54

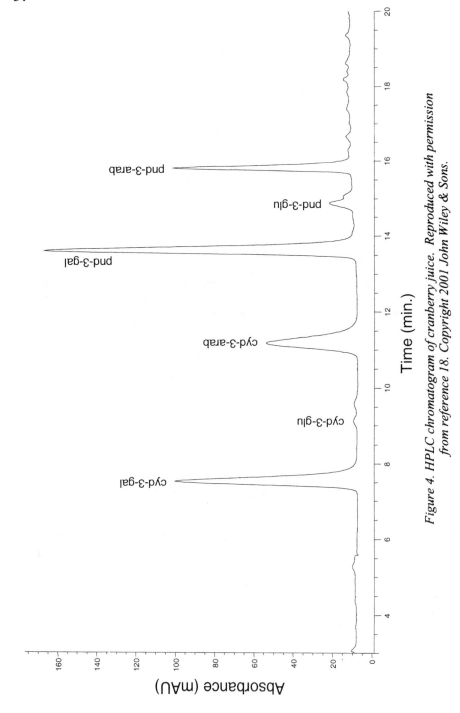

Figure 4. HPLC chromatogram of cranberry juice. Reproduced with permission from reference 18. Copyright 2001 John Wiley & Sons.

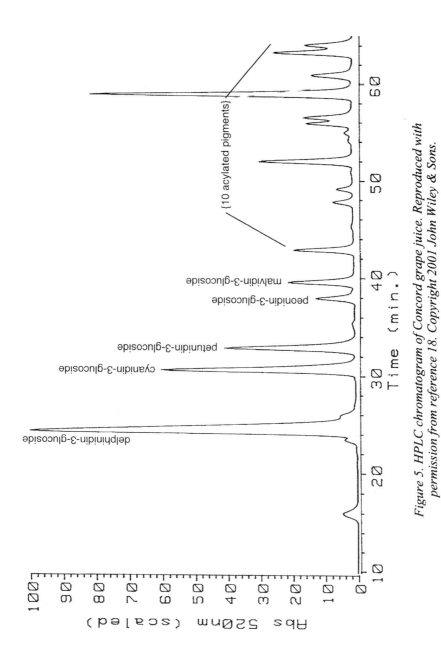

Figure 5. HPLC chromatogram of Concord grape juice. Reproduced with permission from reference 18. Copyright 2001 John Wiley & Sons.

however, offers several advantages for anthocyanin analysis since it provides complete uv-visible spectra for the individual peaks as they are eluted. Considerable information useful for pigment identification can be obtained from the spectra including the wavelength of maximum absorption and the $E_{440\ nm}$ / $E_{vis\ max}$. These measurements help to determine the degree of OH and OCH_3 substitution in the B-ring and whether there is 3 or 3-5 glycosidic substitution (18,35). In addition, the degree and type of cinnamic acid acylation can be inferred from the uv spectrum. Peak purity can also be ascertained by monitoring the spectrum of the peak as it is eluted. Co-elution of polyphenolics can confound interpretation of spectra, but this can be circumvented by isolating the anthocyanins free from other phenolics by solid-phase extraction (18) or diethyl ether precipitation (35).

Reference Standards

Systematic methods are in place for identifying anthocyanin pigments, the work of the analyst being to isolate and purify individual compounds, determine their uv-visible spectral properties, subject them to acid hydrolysis and identify the aglycons and glycosidic substituents, and cleave and identify any acyl substituents by saponification (21,27,35,36). All of this can be a very time-consuming task, but for most anthocyanin-containing commercial materials, the difficult work has been done. Most of the anthocyanins in materials of commercial interest have been identified and are listed in key reference books (11, 38,39). For many investigations the task will be to make peak assignments for previously identified materials. While purified reference standards are becoming more available from commercial firms, the number of compounds is somewhat limited and they can be quite expensive. Retention indices and uv-visible spectra can be obtained from materials which are not purified. Thus extracts of plants and such common materials as cranberry juice cocktail, concord grape juice and anthocyanin-based natural colorants can be used as reference sources (1,40).

Auxiliary Analyses of Hydrolyzed Anthocyanins

Either purified anthocyanins or anthocyanin extracts can be subjected to hydrolysis with 2N HCl to generate the aglycons (18,21,35,36). Reaction conditions are typically 1/2 hr at 100°C. (Note: Anthocyanins acylated with cinnamic acids are resistant to acid hydrolysis requiring much longer hydrolysis times; for acylated pigments it is advisable to first remove acylating acids by the saponification reaction described below and then conduct acid hydrolysis.) The reaction vessel should be flushed with nitrogen and the product should be immediately cooled in an ice bath for the anthocyanidins are extremely unstable. The anthocyanidins are isolated by solid-phase extraction and immediately analyzed. Since there are only six common anthocyanidins, they can be readily be separated and identified by HPLC (*35,36*). The sugars are also liberated and they are best identified by GLC of their tri-methyl silyl derivatives (*36*). Grapes can serve as a source for 5 of the anthocyanidins and strawberries are a good source for pelargonidin, the remaining anthocyanidin. Identification of the aglycons is useful for identifying peaks and for checking authenticity (*1*).

Acylated anthocyanins are readily saponified by treating the isolated anthocyanins with 10% aqueous KOH (*18,35*). The reaction is complete in 8-10 minutes at room temperature in the dark. After neutralization with HCl, the pigments are isolated by solid-phase extraction and then subjected to HPLC analysis. The acylated acids cleaved by saponification can also be isolated, separated by HPLC and characterized from their retention behavior and spectra.

Additional Analytical Methods for Anthocyanin Identification

Mass Spectroscopy of Anthocyanins

Electrospray (ESMS) and tandem mass spectroscopy (MSMS) are powerful techniques for characterization of anthocyanins (*41*). Since anthocyanins are positively charged in an acid environment, soft ionization can produce intact ions, even though the pigments are thermally labile, nonvolatile and polar. Discrete mass units are generated which are extremely useful in confirming peak identities. Purified individual anthocyanins may be introduced by HPLC-ESMS, or alternatively, extracts or mixtures of anthocyanins can be directly injected into the mass spectrometer. Either aqueous or methanolic solutions can be used. At the appropriate voltage, there is little interference from polyphenolics or other neutral compounds because of anthocyanin's positive charge. Cleaner spectra are usually obtained, however, if the anthocyanins are separated from the polyphenolics by clean-up with C_{18} solid-phase extraction using ethyl acetate (*16,18,37,41*). In most cases, cleavage of the glycosidic groups will occur to generate small amounts of the aglycon in addition to the intact anthocyanin. Figure 6 shows the ESMS spectrum for Concord grape (*Vitis labrusca*) anthocyanins. The five aglycons are masses a-e, while 1-5 are the monoglucosides, 6-10 the monoglucosides acylated with *p*-coumaric acid, and 11-15 the di-glucosides acylated with *p*-coumaric acid. Tandem MS-MS provides clear and characteristic fragmentation patterns. Individual molecules are selected by the first quadrupole mass analyzer and fragmented in the collision cell using a suitable gas, usually argon. The fragments are detected by the second quadrupole mass analyzer. This technique will readily distinguish between 3- and 3-5-glycosidic substitution. Figure 7 shows the fragmentation pattern for pelargonidin-3-sophoroside-5-glucoside and delphinidin-3-xylosyl-glucoside. Fragmentation may also identify which sugar substituent contains a given acyl substituent.

58

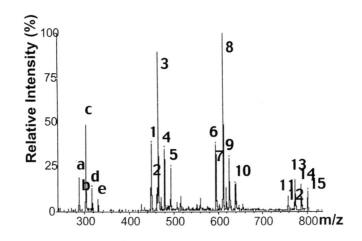

Figure 6. ESMS spectrum of Concord grape juice concentrate anthocyanins. Mass assignments are as follows: (a) cyanidin, (b) peonidin (c) delphinidin (d) petunidin (e) malvidin; (1-5), their respective 3-glucosides; (6-10), their respective 3-glucosides acylated with p-coumaric aid; (11-15), their respective diglucosides acylated with p-coumaric acid. Adapted with permission from reference 41. Copyright 1999 ACS.

MSMS of Pg-3-soph-5-glu

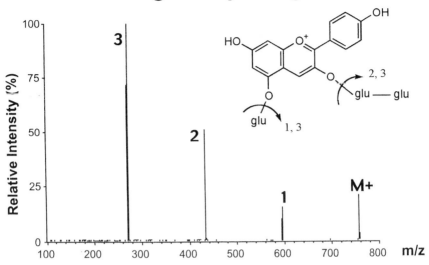

MSMS of Dpd-3-xyl-glu

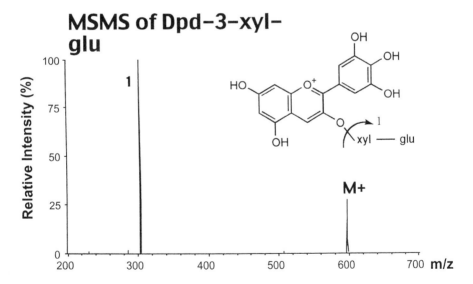

Figure 7. MS-MS fragmentation pattern of anthocyanins pelargonidin-3-sophoroside-5-glucoside (isolated from radish) and delphinidin-3-xylosylglucoside (isolated from Roselle). Reproduced with permission from reference 41. Copyright 1999 ACS.

NMR Spectroscopy of Anthocyanins

Different one-and two-dimension NMR techniques can be used for more complete identification of anthocyanins (*42,43*). Such information as the angle of the glycosidic linkage, sugar identification, location of glycosidic substitution, and site of acylation on the sugar molecule can be determined. In addition conformational information concerning the proximity of neighboring groups can be learned. A limitation of this powerful technique is that mg quantities of highly purified pigments are required to get meaningful data.

What is needed?

This chapter discussed analytical methods which can effectively be used to measure the anthocyanin content and composition of anthocyanin-containing nutraceuticals, foods, and their sources. Some of the immediate industry needs are for standardized nomenclature and standardized protocols for measuring anthocyanin content. Recommended extinction coefficients, and units for expressing anthocyanin pigment content are but two examples which need to be addressed. There is intense research interest in the health benefits of anthocyanins and much to be learned with respect to how, where, and to what extent anthocyanins are absorbed. Additional topics include structure relationships to antioxidant activity, identification of anthocyanin metabolites, determination of relationships between molecular structure and biological activity, determining the specific mechanisms by which anthocyanins may prevent disease, etc. Such research problems require analysis of labile compounds which are present in trace quantities in complex biological matrices. Such are the challenges for the analytical chemist.

References

1. Wrolstad, R. E.; Hong, V.; Boyles, M.J.; Durst, R.W. In *Methods to Detect Adulteration of Fruit Juice Beverages*, S. Nagy, R. L. Wade, Eds., AgScience Inc., Auburndale, FL, **1995**, 260-286.
2. Wrolstad, R.E. In *Natural Food Colorants*, G.J. Lauro, F.J. Francis, Eds., Marcel Dekker, Inc., **2000**, 237-252.
3. Burns, J.; Gartner, P. T.; O'Neil, J.; Crawford, S.; Morecroft, I.; McPhail, D.B.; Lister, C.; Matthews, D.; MacLean, M. R.; Lean, M.E.J.; Duthie, G.G.; Crozier, A. *J. Agric. Food. Chem.* **2000**, *48*, 220-230.
4. Gasiorowski, K.; Szyba, K.; Brokos, B.; Kolaczynska, B.; Jankowiak-Wlodarczyk, M.; Oszmianski, J. *Cancer Letters* **1997**, *119*, 37-46.

5. Boniface, R.; Miskulin, M.; Robert, L.; Robert, A.M. In *Flavonoids and Bioflavonids*, L. Farkas, M. Gabor, F. Kallay, Eds., Elsevier, Amsterdam, **1985**, 293-301.

6. Kajimoto, O. **1999**. Food Style (English translation). *3*, 30-35.

7. Stahelin, H. B. *J. Vitam. Nutr. Res.* **1999**, *69*, 146-149.

8. Wang, H.; Cao, G.; Prior, R.L. *J. Agric. Food Chem.* **1997**, *45*, 304-309.

9. Prior, R.L.; Cao, G.; Martin, A.; Sofic, E.; McEwen, J.; O'Brien, C.; Lischner, N.; Ehlenfeldt, M.; Kalt, W.; Krewer, G.; Mainland, C.M. *J. Agric. Food Chem.* **1998**, *46*, 2686-2693.

10. Kalt, W.; McDonald, J.E.; Donner, H. *J. Food Sci.* **2000**, *65*, 390-393.

11. Strack, D.; Wray, V. In *The Flavonoids, Advances in Research since 1986*, H. B. Harborne, Ed., Chapman and Hall, NY, **1994**, 1-22.

12. Bassa, Y.; Francis, F.J. *J. Food Sci.* **1987**, *52*, 753-754.

13. Rodriguez-Saona, L.E.; Wrolstad, R. E. In *Current Protocols in Food Analytical Chemistry*, R. E. Wrolstad, S. J. Schwartz, Eds, John Wiley and Sons, Inc., NY, **2001**, Unit F1.1.1-F1.1.10.

14. Giusti, M. M.; Wrolstad, R. E. *J. Food Sci.* **1996**, *61*, 322-326.

15. Rodriguez-Saona, L. E.; Giusti, M.M.; Wrolstad, R. E. *J. Food Sci.* **1998**, *63*, 458-465.

16. Skrede, G.; Wrolstad, R. E.; Durst, R. W. *J. Food Sci.* **2000**, *65*, 357-364.

17. Giusti, M. M.; Wrolstad, R. E. In *Current Protocols in Food Analytical Chemistry*, R. E. Wrolstad, S. J. Schwartz, Eds, John Wiley and Sons, Inc., NY, **2001**, Unit F1.2.1-F1.2.13.

18. Durst, R. W.; Wrolstad, R. E. In *Current Protocols in Food Analytical Chemistry*, R. E. Wrolstad, S. J. Schwartz, Eds, John Wiley and Sons, Inc., NY, **2001**, Unit F1.3.1-F13.

19. Wrolstad, R. E.; Culbertson, J. D.; Cornwell, C. J.; Mattick, L. R. *J. Assoc. Off. Anal. Chem.* **1982**, *65*, 1417-1423.

20. Fuleki, T.; Francis, F. J. *J. Food Sci.* **1968**, *33*, 78-82.

21. Francis, F. J. In *Anthocyanins as Food Colors*, P. Markakis, Ed., Academic Press, NY, **1982**, 181-207.

22. Ribereau-Gayon, P. **1959**. Recherches sur les anthocyannes des vegetaux. Application au genre *Vitis*. Ph.D. thesis. University of Bordeaux. Libraire Generale de l'Ensignement, Paris.

23. Jurd, L.; Asen, S. *Phytochemistry* **1966**, *5*, 1263-1271.

24. McClure, J.W. *Plant Phys.* **1967**, *43*, 193-200.

25. Siegelman, H.W.; Hendricks, S.B. *Plant Phys.* **1958**, *33*, 409-413.

26. Zapsalis, C.; Francis, F.J. *J. Food Sci.* **1965**, *30*, 396-399.

27. Swain, T. In *Chemistry and Biochemistry of Plant Pigments*, T. W. Goodwin, Ed., Academic Press, London, **1965**, 533-549.

28. Schou, S.A. *Helv. Chim. Acta.* **1927**, *10*, 907-915.

29. Niketic-Aleksic, G.; Hrazdina, G. *Lebensm. Wiss. U. Technol.* **1972**, *5*, 163-165.

62

30. Somers, T.C. *J. Sci. Food Agric.* **1966,** *17*, 215-219.
31. Koeppen, B.H; Basson, D.S. *Phytochemistry* **1966,** *5*, 183-187.
32. Giusti, M. M.; Rodriguez-Saona, L.E.; Wrolstad, R.E. *J. Agric. Food Chem.* **1999,** *47*, 4631-4637.
33. Jorgensen, E.C.; Geissman, T.A. *Biochem. Biophys.* **1955,** *55*, 389-402.
34. Bakker, J.; Bridle, P.; Honda, T.; Kuwano, H.; Saito, N.; Terahara, N.; Timberlake, C.F. *Phytochemistry* **1997,** *44*, 1375-1382.
35. Hong, V.; Wrolstad, R. E. *J. Agric. Food Chem.* **1990,** *38*, 708-715.
36. Gao, L.; Mazza, G. *J. Agric. Food Chem.* **1994,** *42*, 118-125.
37. Oszmianski, J.; Lee, C.Y. *Am. J. Enol. Vitic.* **1990,** *41*, 204-206.
38. Mazza, G.; Miniati, E. *Anthocyanins in Fruits, Vegetables, and Grains*; CRC Press, Boca Raton, FL, 1993.
39. Macheix, J.J.; Fleuriet, A.; Billot, J. *Fruit Phenolics*. CRC Press, Boca Raton, FL, 1990.
40. Hong, V.; Wrolstad, R. E. *J. Agric. Food Chem.* **1990,** *38*, 698-708.
41. Giusti, M. M.; Rodriguez-Saona, L. E.; Griffin, D.; Wrolstad, R. E. *J. Agric. Food Chem.* **1999,** *47*, 4657-4664.
42. Giusti, M. M.; Ghanadan, H.; Wrolstad, R. E. *J. Agric. Food Chem.* **1998,** *46*, 4858-4863.
43. Fossen, T.; Andersen, O.M.; Ovstedal, D.O.; Pedersen, A. T.; Raknes, A. *J. Food Sci.* **1996,** *61*, 703-706.

Chapter 5

Role of Marker Compounds of Herbs and Their Bioavailability in Quality, Efficacy, and Safety of Products

Harunobu Amagase

Department of Research and Development, Wakunaga of America Company, Ltd., 23501 Madero, Mission Viejo, CA 92691

While the herbal supplement market has grown rapidly in recent years, the public is heavily scrutinizing the quality of available products. Although the products should be controlled by quality control systems, including standardization with marker compounds based upon legitimate science to ensure consistent high quality as well as valid efficacy and safety, there is still a great deal either unknown or misunderstood regarding the actual active compounds of herbs. Bioavailability is one of the issues often ignored for standardization of herbs due to the complexity of chemicals, even though it is essential for the utility of active ingredient(s) and is critical for assuring product quality. It is also beneficial for confirming the compliance of the subjects in clinical trials.

Herbal Market and Current Confusion

The herbal supplement market has grown rapidly in recent years, especially since the Dietary Supplement Health and Education Act (DSHEA) was passed in

1994. According to a phase one survey of vitamins, minerals, herbs, and supplements conducted in 1997, garlic (*Allium sativum*) supplements were number one in sales among 91 herbal dietary supplements (*1*). The herbal supplement market is steadily growing, but the public still heavily scrutinizes this industry (*2-4*).

Several clinical study reports of garlic supplementation, including meta-analyses have revealed a cholesterol-lowering effect in humans (*5-8*). These studies have had a strong impact on public awareness of the cholesterol lowering effect of garlic. However, recent publications (*9-13*) have reported that a garlic oil product and a dehydrated garlic powder product have no effect on cholesterol levels in human plasma. These negative studies, in light of the many positive ones on garlic in general, are likely the main cause of confusion promulgated in the general public and academic societies through the media (*14*). The most recent meta-analysis of randomized clinical trials on dehydrated garlic powder products that excluded aged garlic extract (*15*) concluded the use of garlic powder for hypercholesterolemia is of questionable value. These negative publications have also caused sales to drop and have generated a bad image for the industry as a whole. Sales of garlic supplements have dropped significantly in recent years, dropping 28 percent, from \$223 million in 1997 to \$174 million in 1999 (*16,17*).

Borrowing science is another issue. Only two garlic supplement manufacturing companies have researched their own products in scientific studies including clinical trials. Many companies, including major players in the market, are not performing any scientific studies. Therefore, these companies just borrow the published literature from other companies even though their preparations or product characteristics are completely different. One company utilizes a Swiss patent and create a unique formula with effects found in general garlic but it does not have any clinical studies. Although several companies manufacture enteric-coated (delayed-release) garlic tablets, these products have not been clinically tested for either efficacy or safety.

Misunderstanding Garlic Chemistry

One of the main reasons for the negative results about garlic powder products may be standardization with the wrong marker compound, allicin or allicin potential/yield. Since allicin is:

1. not present in garlic or any garlic supplements,
2. not bioavailable,
3. not the active ingredient in garlic or
4. not a good marker compound for standardization of quality control,

using allicin or allicin potential to standardize a product creates inconsistent results in clinical studies. Freeman reported that allicin is not present in any garlic products in the market and allicin is not generated in the stomach (18). Although the active ingredient of garlic is not known at this moment and since garlic has more than 200 kinds of compounds, including various sulfur compounds (19), standardization in these studies should be done with compounds that are known to be at least bioavailable and effective.

Allicin has been believed to be the active ingredient in garlic for so long simply because it bears the odor of garlic. Although the notorious odor of garlic is attributed to allicin and other oil-soluble sulfur components generated by crushing the intact bulb, the number of these sulfur components is limited. Once garlic is processed by cutting or crushing, preserved compounds in the intact garlic are almost immediately converted to hundreds of degraded organosulfur compounds by a cascade of enzymatic reactions. The primary sulfur-containing constituents, γ-glutamylcysteines and S-alkyl-L-cysteine sulfoxides, including alliin, are abundant in the intact cloves. The γ-glutamyl peptides are biosynthetic intermediates for corresponding cysteine sulfoxides (20) and are the reserved compounds that can be hydrolyzed and oxidized to form alliin. When garlic is "damaged", i.e. attacked by a microbe, crushed, cut, chewed, dehydrated, or pulverized and exposed to water, the vacuolar enzyme, alliinase, rapidly lyses the cytosolic cysteine sulfoxides (alliin) to form cytotoxic and odiferous alkyl alkane-thiosulfinates, such as allicin, and other transiently formed compounds. Allicin comprises 70-80% of these thiosulfinates and is an oily colorless liquid that exists fleetingly, being quickly reduced to various degradation compounds. It has also been shown that no allicin can be detected in the blood after the ingestion of raw garlic juice containing high amount of allicin (21). These findings clearly indicate that allicin itself cannot contribute to any of the beneficial effects of garlic inside the body. Currently, allicin is thought to be a transient compound that is rapidly decomposed into many different kinds of sulfur compounds.

The intact garlic bulb also contains small amounts of S-allyl cysteine (SAC), one of the water-soluble sulfur compounds derived from garlic. SAC is formed along another pathway than that taken for alliin to be converted to allicin. However, despite this it contributes to the health benefits of garlic. SAC is the most widely studied constituent in garlic and its biological activities have been shown through many scientific publications (22-27).

As additional constituents, the following compounds are present in intact garlic: Steroidal glycosides (28), lectins (29), prostaglandins, fructan, pectin, essential oil, adenosine, vitamins-B_1, -B_2, -B_6, -C, and -E, biotin, nicotinic acid, fatty acids, glycolipids, phospholipids, anthocyanins, flavonoids, phenolics, and essential amino acids (*30*). All these components act synergistically, but the amounts vary by the method of preparation.

Garlic Products on the Market

Garlic products have enjoyed increasing popularity in the last decade. They provide an opportunity to conveniently share in the health benefits of garlic. Sulfur-containing compounds in commercial garlic preparations vary depending on their manufacturing processes. There are dozens of brands of garlic on store shelves, however, they are classified into only four groups, such as garlic essential oil, garlic oil macerate, garlic powder and garlic extract. Efficacy and safety of a garlic supplement is contingent upon its manufacturing process. The chemistry of garlic is quite complicated as stated above, and different types of processing produce more than just preparations in different forms. These forms also differ in their chemical ingredients, efficacy and toxicity. Documentation of safety and effectiveness are crucial in the evaluation of products used for health purposes, such as drugs and food supplements. Garlic products that contain the most safe, effective, stable, and odorless components are the most valuable and practical as food supplements.

It is well known that extraction increases potency and bioavailability of various crude botanicals including garlic, and eliminates harsh and toxic characteristics. According to many studies on one of the extracts, aged garlic extract, results in greater efficacy and safety compared to raw garlic, dehydrated garlic powder or other garlic preparations.

Essential Oil

Garlic essential oil is obtained by steam distillation of garlic. Ground whole garlic cloves in water are heat-distilled or extracted in an organic solvent (e.g., hexane) to obtain the oil fraction. Water-soluble compounds, which are reputed to be effective, are totally eliminated by this process. The essential oil content in cloves is 0.2 to 0.5 % and consists of a variety of sulfides, such as diallyl disulfide and diallyl trisulfide (*31-32*). Allicin is also completely eliminated from oil. Most of these commercially available capsules contain mainly vegetable oil and a small amount of garlic essential oil because its pungent odor causes bad breath. A recent publication on garlic oil (*9*) revealed that garlic oil was not effective on cholesterol lowering in humans. Generally garlic oil has very poor results on cholesterol.

Dehydrated Powder

Garlic powder is mass-produced as a flavoring powder for condiments and processed foods. Garlic cloves are sliced or crushed, dried and pulverized into powder. Garlic powder is thought to retain the same ingredients as raw garlic; however, the proportions are quite different than that of raw garlic. Dehydrated powders contain similar constituents as garlic cloves, but they differ significantly in amounts of these constituents (33). For example, the main sulfur compound in both raw garlic and garlic powder is alliin. A pure dehydration process losing no ingredients results in an alliin content in the powder of 2-2.5%. However, garlic powders only comprise 1% of alliin at most, meaning that more than half of the alliin is lost during processing. Therefore, garlic powders may contain similar constituents as raw garlic, but their proportions are not the same and their content varies depending on the dehydrating conditions. While allicin is often emphasized in dehydrated powder, all preparations tested contained no allicin due to its instability (18,34,35).

Although some garlic powder products claim to generate a certain amount of allicin (the so-called "allicin potential" or "allicin yield"), only a very small amount of allicin was produced in a simulated gastric solution (USP method) compared to water (18), demonstrating that allicin can not be generated in our stomach from garlic powder products. Therefore, allicin is neither the active compound nor an appropriate marker compound in garlic. Standardization using this compound is not scientifically reasonable. Further, standardization for ingredients, safety, and efficacy are not well established for dehydrated garlic powders, especially enteric-coated (or delayed-release) products. Actually, nobody has found allicin in the blood after consumption of any garlic preparation, including raw garlic and enteric-coated garlic products.

Oil Macerate

This garlic product was originally developed for use as a condiment. Whole garlic cloves are ground in a vegetable oil and the mixture is encapsulated. During the manufacturing process, some amount of alliin is converted to allicin. Since allicin is unstable and decomposes quickly, oil macerate preparations contain allicin-decomposition compounds, such as dithiins, ajoene and sulfides, residual amount of alliin and other constituents in garlic (31,36). Standardization for ingredients, safety, and efficacy are not well established for oil macerates.

Extract

Whole or sliced garlic cloves are soaked in an extracting solution (e.g., purified water and diluted alcohol) for a certain period of time. After separating the solution, the extract is generally concentrated and used for preparations. A powder form of extract is also available. The extract, especially aged garlic extract, contains mainly water-soluble constituents and a small amount of oil-soluble compounds (*37*). Namely, aged garlic extract is characterized by water-soluble sulfur-containing compounds including SAC and *S*-allyl mercaptocysteine (*24*). Aged garlic extract is processed in a different way from the other three types of garlic products. As the name indicates, this extract is extracted for up to twenty months (aged). During this aging process, odorous, harsh, and irritating compounds in garlic are naturally converted into stable and safe compounds. It is standardized with SAC (USP Garlic Fluidextract monograph) and has an array of efficacy and safety studies.

Bioavailability of Garlic Compounds

Bioavailability of active ingredients is essential. SAC is one of the water-soluble organosulfur compounds in garlic and its concentration increases by the progression of extraction/aging. The pharmacokinetics of SAC is well established in several animals (*38*) and confirmed in humans (*39*). SAC is detected in human blood and its blood concentration and other pharmacokinetic parameters are strongly associated with dosages of orally administered SAC (*38,39*). N-acetyl-SAC is a metabolite of SAC found in human urine after garlic consumption (*40,41*). This indicates that SAC could be transformed to form the N-acetylated metabolite by N-acetyltransferase in the body. The bioavailability of SAC is 103.0% in mice, 98.2% in rats and 87.2% in dogs (*38*). SAC and its metabolite(s) are currently the only compliance maker compounds for clinical human studies involving garlic consumption (*39*). Therefore, since SAC is present in most of the garlic preparations, has many biological effects and is bioavailable, it should be considered one of the active principles in garlic preparations. Thus, the standardization of garlic using SAC is scientifically reasonable and well justified.

The oil-soluble organosulfur compounds in garlic, including allicin, sulfides, ajoene and vinyldithiins, were not found in human blood or urine from 1 to 24 hours after ingestion of 25 g of raw garlic containing a significant amount of allicin (*21*). A preliminary clinical trial of the commercially available garlic products, including enteric-coated products, has revealed no allicin in human blood following consumption. When allicin was perfused into isolated rat livers, it showed a remarkable first pass effect and metabolized to diallyl disulfide and allyl mercaptan. Moreover, allicin disappeared very rapidly when incubated with liver homogenate (42). Seventy percent of the radioactivity was distributed in the liver cytosol, of which 80% was metabolized to sulfate. However, this does not indicate the bioavailability of allicin, since radioactivity is determined from the sulfur atom, not the entire compound.

Comparing the *in vitro* anti-platelet aggregating effects of garlic products on the market, based on their ajoene and dithiin concentration (21) can be misleading and inaccurate without biological tests to confirm the bioavailability and efficacy of these compounds *in vivo*. For example, aged garlic extract does not contain either of these compounds but has shown significant reduction in platelet aggregation and adhesion in randomized, double-blind, placebo-controlled clinical studies (43,44). Since ajoene and dithiin are not detectable in the blood, testing these compounds *in vitro* is meaningless. Due to the lack of bioavailability, they do not contact blood cells. On the other hand, based upon the above evidence, SAC and its metabolite(s) are currently the only compliance maker compounds for clinical human studies involving garlic consumption (39-41). Furthermore, SAC and its metabolites can be active compounds in the body.

Other metabolites of garlic constituents, such as *N*-acetyl-*S*-(2-carboxypropyl)-cysteine and hexahydrohippuric acid, have been detected in human urine after ingestion of garlic (40).

The active principles in garlic have not been fully identified, but bioavailable sulfur-containing compounds, such as SAC must play an important role in the pharmacological activities of garlic. Pharmacological effects of SAC are quite widespread as shown both *in vitro* and *in vivo*. SAC inhibits cholesterol synthesis in hepatocyte cultures (45) and in chickens (46). It increases glutathione level in various cells (47,48), inhibits carcinogen binding to mammary DNA (23), possesses antioxidative effects (24), brain-protective effects (26,49,50), hepato-protective effects against various chemicals (47,51) and other effects. Since it is bioavailable and can reach specific organs and cells, it is reasonable to study SAC in *in vitro* systems.

Other Active Ingredients and Wide Variety of Efficacy of Garlic

The optimal effect of aged garlic extract on peripheral microcirculation in healthy human subjects occurs around 90-120 min (52), the same time when the optimal SAC blood level is obtained in the pharmacokinetics study (38). However, other effects of SAC occur much later. The blood pressure lowering effect of the aged garlic extract happens several days after consumption (53).

Inhibition of platelet aggregation and adhesion by the aged garlic extract requires about 2 weeks to 3 months (43). Cholesterol-lowering effects in the clinical studies were observed after 3-6 months of consumption (5,53,54).

In addition, the concentration of SAC at the target site in in vitro systems and clinical studies are quite different. There is a big gap between them and the usual concentration in the in vitro system is much higher than that of human subjects. This leads to the following hypothesis:

1. there must be other active ingredients in the garlic/garlic preparations in addition to SAC,
2. these various compounds must work synergistically, and
3. SAC must work by multiple mechanisms and stimulate various cascades of actions.

Furthermore, garlic improves not only cardiovascular health, but it has also demonstrated anticancer, immune, detoxification, antioxidative, antistress and antiaging effects. Even within the cardiovascular area, the clinical effects of garlic are widespread on serum cholesterol, triglycerides, blood viscosity, platelet functions, vasodilation (microcirculation), blood pressure, stiffness of artery (elasticity), LDL oxidation, sickle cell anemia, recovery of hematological figure, anemia related dizziness or lightheadedness, hand and foot coldness, flutter (heart palpitations), lumbago, shortness of breath and others.

Thus, many active constituents in garlic may influence the various effects of garlic in the cardiovascular area. Furthermore, though milder than drug activity, garlic can cover a much wider range of pharmacology, reducing a varied number of risk factors in a moderate way. On the contrary, synthesized drugs are very specific in their actions and tend to inhibit only very specific risk factors of a disease.

Issues of Markers in Other Herbals

Flavonoids and Intestinal Flora

Flavonoids are among the most interesting phytochemicals in herbs, especially in *Ginkgo biloba*, soy and others. The importance of intestinal bacteria for absorption and metabolism of flavonoids is well known. When ingested, various forms of flavonoids undergo hydrolysis by β-glucosidases from intestinal bacteria in the jejunum, releasing the principle bioactive aglycones (sugar-free metabolites) (55,56). Further metabolism takes place in the consequent intestine with the formation of specific metabolites (57). The aglycones and any bacterial metabolites are absorbed from the intestinal tract and conjugated mainly to glucuronic acids, then they undergo enterohepatic recycling. Intestinal metabolism is essential for subsequent absorption and bioavailability (58) because there is no evidence to support absorption of the conjugated form of isoflavones (59), so that orally ingested larger amounts of flavonoids might be present mainly as aglycones in the intestine, from which they could be absorbed with miscelles of bile acids into the epithelium and then into the blood. Thus, the absorption and blood level of aglycones of flavonoids are dependent upon the condition of the individual intestinal flora.

Hypericin and Stability

Stability of markers is also very important for herbal products. Hypericin in St. John's Wort is well known as its marker compound and believed to be active for a long time. However, since it is very unstable, in the US Pharamcopoeia (USP) hyperiforin is employed as an additional chemical marker for the standardization of St. John's Wort. Furthermore, bioavailability of active or marker compounds is another issue to be resolved in the case of St. John's Wort. This issue was heavily discussed in the *3rd International Congress on Phytomedicine* in Munich (60).

Overall Issues/Analytical Methods and Reference Standard

Established analytical methods and high quality reference standards are urgently necessary for the purpose of quality control of herbal products. USP is currently organizing herbal monographs under National Formula (NF) and establishing analytical methods and reference standards for their analyses based upon updated scientific information. These movements are definitely essential for high quality herbal products to remain in the market.

Quality Control and Regulatory Issues

The effectiveness of herbs is in the prevention of health problems rather than therapy. To obtain the preventative benefits of herbs, long term supplementation is necessary. With long term supplementation, the issue of toxicity/safety must be carefully considered. Documentation of safety is crucial in the evaluation of products used for health purposes, such as drugs and food supplements. As mentioned previously, different garlic preparations comprise different constituents, which necessitates toxicological tests to ensure their safety. Thus, the safety of specific garlic preparations should be a major consideration in quality control. Contraindication with medication should also be carefully investigated. In case of aged garlic extract (Kyolic®), no contraindication with Coumadin® (warfarin) was presented (61), but other forms of garlic preparations, including raw garlic, may not have similar interactions due to the different chemical characteristics. Scientifically reasonable quality control is essential for high quality standard products, including the various studies, such as safety, contraindication and others. USP and other organizations are now working on the garlic monograph as part of a series of botanical monographs. Although some monographs still have allicin in the text, USP has made a scientific judgment regarding the active ingredient(s) in garlic and removed allicin from its garlic monograph. Since the labeled statements on herbal supplements should be based on reasonable science and compounds which are actually present, this authorized monograph is very important for industries and the public based upon the philosophy of consumer protection.

Difference Between Marker(s) and Active Compound(s)

The marker- and active-compounds of herbal products are not always the same. In the case of garlic, SAC is apparently a good marker compound for standardization of preparations and is one of the active compounds in garlic products. However, it is not the only active ingredient responsible for efficacy, as mentioned above. There are many compounds in herbal preparations that work synergistically so that emphasis should not be put on only one.

Since there are many compounds in herbals, these various compounds must have various mechanisms of action. In addition, different time frames for efficacy, as mentioned above, should also be considered as a difference between markers and actives.

Since these are serious issues for the standardization of herbal products, more pharmacodynamic studies are needed to clarify the details of the difference and accurate determination of actives and markers in herbal products.

Difference Between Herbal Products and Synthesized Drugs

There is also a big difference between herbal products and synthesized chemical drugs. First, chemical constituents in herbals are generally quite complicated, whilst a synthesized chemical drug is a single chemical entity. A single chemical drug itself is the active compound, but usually actives in herbal products are unknown. Efficacy of both preparations is quite different. The degree of efficacy of single chemical entities is quite specific and evident, but that of herbal products is generally vague and it is difficult to determine the exact dosage for specific diseases.

Safety is another concern. Herbal products are generally believed to be safe, but safety is dependent upon the herb in question. The procedure for determining safety is well established for chemically synthesized drugs, but not for herbal supplements.

Specific regulation of herbal products as drugs was recently released by the FDA in the USA (62). Herbal medicines should be regulated under special quality control and the same ordinal pathway as chemically synthesized drug for the same regulation. The quality control standard and procedure are also not established yet. Further consensus among the regulatory office, academia and industries are necessary to distribute high quality and reliable herbal products.

The above situation of herbal products, especially their difference compared to chemically synthesized drugs, is basically dependent upon the cultural background of registration of herbal drugs. Since it is quite difficult for the USA to register multiple herbal combination products, such as traditional Chinese medicines, these combined herbal products generally require a longer process and case by case judgment for approval.

Conclusion

1. Herbals should be carefully studied, evaluated, and viewed from various standpoints.
2. Markers of herbals should be carefully selected based on reasonable, updated science including studies of pharmacokinetics and bioavailability.
3. Since marker and active compounds are not always the same, more studies are needed to clarify the entire picture of chemistry, efficacy and safety of herbs.

References

1. Wyngate, P. *Natural Foods Merchandiser*. March 1998, p. 14.
2. Moore, T. *Washingtonian.* July 1999.
3. Editorial. N. Engl. J. Med. **1998**, *339 (12)*.
4. *Nutrition Action Healthletter* **2000**, *27(8)*, 8-9.
5. Lau, B.H.S., Lam, F., Wang-Cheng, R. *Nutr. Res.* **1987**, *7*, 139-149.
6. Warshafsky, S., Kamer, R., Sivak, S. *Ann. Intern. Med.* **1993**, *119*, 599-605.
7. Silagy, C., Neil, A. *J. Royal Coll. Physic. Lond.* **1994**, *28*, 39-45.

74

8. Neil, A.; Silagy, C.; Lancaster, T.; Hodgeman, J.; Vos, K.; Moore, J.; Jones, J.; Cahill, J.; Fowler, G. *J. Royal Coll. Physic. Lond.* **1996**, *30*, 329-334.
9. Berthold, HD.; Sudhop, T.; Bergmann, KV. *JAMA* **1988**, *279*, 1900-1902.
10. Isaacsohn, JL.; Moser, M.; Stein, E.; Dudley, K.; Davey, J.; Liskov, E.; Black, H. *Arch Intern. Med.* **1998**, *158*, 1189-1194.
11. Simons, LA.; Balasubramaniam, S.; von, Konigsmark, M.; Parfitt, A.; Simons, J.; Peters, W. *Athlerosclerosis* **1995**, *113*, 219-225.
12. Breithaupt-Grögler, K.; Ling, M.; Boudoulas, H.; Belz, G. *Circulation* **1997**, *96*, 2649-2655.
13. McCrindle, B.; Helden, E.; Conner, W. *Arch. Pediatr. Adolesc. Med.* **1998**, *152*, 1089-1094.
14. Altman, L. No Heart benefits seen in heart trial. *New York Times*. New York, New York, December 25, 1996.
15. Stevinson, C.; Pittler, M.; Ernst, E. *Ann. Intern. Med.* **2000**, *133*, 420-429.
16. Anonymous. *Nutrition Business Journal* **2001**, *7*, March issue.
17. Schorr, M. Garlic stinks? September 19, 2000. *ABC News.com.* (http://abcnews.gol.com/sections/living/DailyNews/garlic000918.html)
18. Freeman, F.; Kodera, Y. *J. Agric. Food Chem.* **1995**, *43*, 2332-2338.
19. Block, E. *Angew. Chem. Int. Ed. Engl.* **1992**, *31*, 1135-1178.
20. Lancaster, J.; Shaw, M. *Phytochemistry* **1989**, *28*, 455-460.
21. Lawson, L.; Ransom, D.; Hughes, B. *Throm. Res.* **1992**, *65*, 141-156.
22. Lee, E.; Steiner, M.; Lin, R. *Biophys. Acta* **1994**, *1221*, 73-77.
23. Amagase, H.; Milner, J. *Carcinogenesis* **1993**, *14*, 1627-1631.
24. Imai, J.; Ide, N.; Nagae, S.; Moriguchi, T.; Matsuura, H.; Itakura Y. *Planta Med.* **1994**, *60*, 417-420.
25. Li, G.; Qiao, C.; Lin, R.; Pinto, J.; Osborne, M.; Tiwari R. *Oncol. Reports* **1995**, *2*, 787-791.
26. Numagami, Y.; Sato, S.; Ohnishi, T. *Neurochem. Int.* **1996**, *29(20)*, 135-143.
27. Sumiyoshi, H.; Wargovich, M. *Cancer Res.* **1990**, *50*, 5084-5087.
28. Matsuura, H; Ushiroguchi, T; Itakura, Y; Hayashi, H; Fuwa, T. *Chem. Pharm. Bull.* **1989**, *37*, 2741-2743.
29. Kaku, H.; Goldstein, I.; Van Damme, E.; Peumans, W. *Carbohydrate Res.* **1992**, *229*, 347-353.
30. Fenwick, G.; Hanley, A. *Crit. Rev. Food Sci. Nutr.* **1985**, *22(4)*, 273-377.
31. Block, E. *Sci. Am.* **1985**, *252(3)*, 94-99.
32. Yan, X.; Wang, Z.; Barlow, P. *Food Chem.* **1992**, *45*, 135-139.
33. Iberl, B.; Winkler, G.; Müller, B. *Planta Med.* **1990**, *56*, 320-326.
34. Yan, X.; Wang, Z.; Barlow, P. *Food Chem.* **1993**, *47*, 289-294.
35. Rosen, R. *Phytomed.* **2000**, *7(2)*, 51.
36. Iberl, B.; Winkler, G.; Knobloch, K. *Planta Med.* **1990**, *56*, 202-211.

37. Weinberg, D.; Manier, M.; Richardson, M.; Haibach, F. *J. Agric. Food Chem.* **1993**, *41*, 37-41.
38. Nagae, S.; Ushijima, M.; Hatono, S.; Imai, J.; Kasuga, S.; Matsuura, H.; Itakura, Y.; Higashi, Y. *Planta Med.* **1994**, *60*, 214-217.
39. Steiner, M. *J. Nutri.* **2001**, *131* 980S-984S.
40. Jandke, J.; Spiteller, G. *J. Chromatography* **1987**, *421*, 1-8.
41. De Rooij, BM.; Boogaard, P.; Rijksen, D.; Commandeur, J.; Vermeulen, N. *Arch. Toxicol.* **1996**, *70(10)*, 635-639.
42. Egen-Schwind, C.; Eckard, R.; Kemper, F. *Planta Med.* **1992**, *58*, 301-305.
43. Steiner, M.; Lin, R. *J. Cardiovas. Pharmacol.* **1998**, *31*, 904-908.
44. Rahman, K.; Billington, D. *J. Nutr.* **2000**, *130*, 2662-2665.
45. Yeh, Y.; Yeh S. *Lipids* **1994**, *29*, 189-193.
46. Abuirmeileh, N.; Yu, S.; Qureshi, N.; Lin, R.I.-S.; Qureshi, A. *FASEB J.* **1991**, *5:* 8048.
47. Nakagawa, S.; Kasuga, S.; Matsuura, H. *Phytother. Res.* **1988**, *1*, 1-4.
48. Pinto, J.; Qiao, C.; Xing, J.; Rivlin, R.; Protomastro, M.; Weissler, M.; Thaler, H.; Heston, W. *Am J. Clin. Nutr.* **1997**, *66*, 398-405.
49. Nishiyama, N.; Moriguchi, T.; Matsuura, H.; Itakura, Y.; Katsuki, H.; Saito, H. *J. Nutri.* **2001**, *131*, 1093S-1095S.
50. Moriguchi, T.; Matsuura, H.; Kodera, Y.; Itakura, Y.; Katsuki, H.; Saito, H.; Nishiyama, N. *Neurochem. Res.* **1997**, *22*, 1449-1452.
51. Nakagawa, S.; Yoshida, S.; Hirao, Y.; Kusaga, S.; Fuwa, T. *Hiroshima J. Med. Sci.* **1985**, *34,* 303-309.
52. Okuhara, T. *Jpn. Pharmacol. Therapeut.* **1994**, *22*, 3695-3701.
53. Steiner, M.; Kham, A.; Holbert, D.; Lin, R. *Am. J. Clin. Nutr.* **1996**, *64*, 866-870.
54. Yeh, Y.; Lin, R.; Yeh, S. *J. Am. Coll. Nutr.* **1995**, *14*, 545.
55. Brown, J. *Mutat. Res.* **1980**, *75*, 243-277.
56. Tamura, N.; Agrawal D.; Townley R.; Braquet P.; Fecalase. *Proc. Natl. Acad. Sci. USA* **1980**, *77*, 4961-4965.
57. Joannou, G.; Kelly, G.; Reeder, A.; Waring, M.; Nelson, C. *J. Steroid Biochem. Mol. Biol.* **1995**, 54, 167-184.
58. Setchell, K.; Borriello, S.; Hulme, P.; Kirk D.; Alexon, M. *Am. J. Clin. Nutr.* **1984**, *40*, 569-587.
59. Setchell, K. *J. Nutr.* **2000**, *130*, 654S-655S.
60. Bonn, M; Derendorf, H. Pharmacokinetic/Bioavailability Studies with Various Phytopharmaceuticals. 3rd International Congress on Phytomedicine. October 12, **2000**, Munich, Germany.
61. Rozenfeld, V.; Sisca, T.; Callahan, A.; Crain, J. American Society of Health-System Pharmacists, December 3-7, **2000**, Las Vegas, Nevada.
62. *Federal Register*, August 11, **2000**, *65(156)*, 49247.

Chapter 6

Quality Management of Marine Nutraceuticals

Fereidoon Shahidi[1] and Se-Kwon Kim[2]

[1]Department of Biochemistry, Memorial University of Newfoundland,
St. John's, NF A1B 3X9, Canada
[2]Department of Chemistry, Pukyong National University,
Pusan 608–737, Korea

Marine nutraceutical encompass a wide array of products such as fish oils, seal oil, fish liver oils, algal products, bioactive peptides and protein hydrolyzates, chitosan, its oligomers and glucosamine as well as mucopolysaccharides and algal products, among others. The quality of products involved depends heavily on the quality of raw material, methods of processing and preparation as well as storage stability of products. This overview provides an account of important issues involved in the quality management of selected nutraceuticals from marine resources.

Marine nutraceuticals are bioactive products that originate from aquatic species. Such preparations are often used in the pharmaceutical form of capsule, pill, powder or liquid and are involved in health promotion and disease prevention. Although such bioactives are often present as such in the source materials, their content might not be sufficient to exert their beneficial effects or that consumers are reluctant to use them as foods due to their rigid dietary habits or lack of availability of products.

Among nutraceuticals of interest from the aquatic environment, marine oils, chitinous materials and bioactive peptides are most important. However, algal and microalgal species may also serve as a source of a multitude of nutraceuticals. Many of the bioactives may also be extracted from processing discards of commercial fisheries and hence there is a keen interest in producing value-added components for commercial purposes. However, quality management of marine nutraceuticals, similar to those from other resources requires rigid control in order to offer high quality products to the market. As an example, highly unsaturated fatty acids (HUFA) of the omega-3 family that are found abundantly in fatty fish species are highly prone to oxidation. Therefore, processing, packaging and storage of such products and the form in which they are offered to the market should address issues related to the flavor quality of products. To alleviate concerns about off-flavors in products, marine oils are often sold in the capsule form. Nonetheless, for food applications one must address such important issues as the consumers will not buy a product that suffers from flavor defects even though he or she might not be aware of the harmful effects of oxidized products.

This chapter provides an overview of certain quality issues related to selected nutraceuticals from marine resources. In particular, marine oils, chitinous materials, protein hydrolyzates and algal sources will be discussed.

Marine Oils

Marine oils originate from the body of fatty fish species such as herring, mackerel and menhaden as well as liver of white lean fish such as cod and halibut and blubber of marine mammals such as seals and whales. In addition, algal and microalgal species provide an important source of oil which resemble those from other marine resources.

Marine oils contain highly unsaturated fatty acids with multiple unsaturated olefinic groups and these belong to the omega-3 family of fatty acids. The major fatty acids involved include eicosapentaenoic acid (C20:5ω3; EPA), docosahexaenoic acid (C22:6ω3, DHA) and to a lesser extent docosapentaenoic acid (C22:5ω3, DPA) (1). These fatty acids have been demonstrated to provide unique health benefits to consumers. However, the highly unsaturated nature of the molecules involved presents the scientists and technologists with a difficult challenge in delivering them in a form that does not have off-flavors associated with oxidation products of highly unsaturated fatty acids (HUFA).

Interest in omega-3 fatty acids as health-promoting and nutraceutical agents has expanded dramatically in recent years as a rapidly growing body of literature indicates their effects on alleviating cardiovascular disease, type 2 diabetes, inflammatory ailments, and autoimmune disorders, among others. To this end, the oils may be used as such or in the form of enriched omega-3 oils. The latter may be in the free fatty acid, simple alkyl ester or acylglycerol forms.

The quality of marine oils and their concentrates is dictated by a manifold of factors. Some significant issues in this relation are discussed below.

Fatty Acid Composition and Positional Distribution

The fatty acid composition of marine oils reveals the presence of different proportions of omega-3 fatty acids in them. Among these the ratio of DHA to EPA and possible presence of DPA in modest amounts are most important. While DHA is involved in tissues with electrical activity, namely the heart, the eye and the brain, EPA has its own benefits with regard to its participation in eicosanoid synthesis. In pregnant and lactating women and in infants DHA is very important as it is an integral part of the brain and retina and constitutes approximately 50% of the fatty acids in the phospholipids involved and hence participates in their development and health status (2). For cardiovascular disease, autoimmune disorder and type 2 diabetes, both EPA and DHA are known to serve important roles (3). The proportion of EPA, DPA and DHA in different marine oils is provided in Table I.

It is interesting to note that DPA is present in significant amounts only in seal blubber oil (1). However, presence of $C22:5\omega3$ in certain algal oils has also been reported (4). The importance of DPA of the omega-3 family in health promotion has not yet been adequately studied and investigations in this area are warranted. However, DPA may act as a powerful anti-atherogenic factor (5).

The presence and distribution of EPA, DPA and DHA in the triacylglycerol molecules is dictated by the source material (Table II). However, little has so far been done to examine the dependence of health benefits of marine oils on positional distribution of omega-3 fatty acids involved. However, it has been observed that dominance of EPA, DPA and DHA in the sn-1 and sn-3 is more helpful in alleviating inflammatory swelling and arthritic pain of the joints.

In omega-3 concentrates, the form of the preparation is another important factor to be considered. In general, for health promotion and as nutraceuticals, acylglycerols are preferred. However, for clinical and therapeutic applications, use of concentrates with a very high proportion of omega-3 fatty acids may be necessary. Thus, concentrates of omega-3 fatty acids may be prepared in the free fatty acid or simple alkyl (methyl and ethyl) ester forms (6, 7).

Table I. Major Fatty Acids (>2%) of Marine and Algal Oils (w/w %).

Fatty Acid	Menhaden Oil	Cod Liver Oil	Seal Blubber Oil	Algal Oil[a]
14:0	8.32	3.33	3.73	14.9
16:0	17.1	11.1	5.98	9.05
16:1 ω7	11.4	7.85	18.0	2.20
17:0	2.45	0.61	0.92	–
18:0	3.33	3.89	0.88	0.20
18:1ω9	8.68	16.6	20.8	18.9
18:1ω11	3.46	4.56	5.22	–
18:4ω3	2.90	0.61	1.00	–
20:1ω9	1.44	10.4	12.2	–
20:5ω3	13.2	11.2	6.41	–
22:1ω11	0.12	9.07	2.01	–
22:5ω3	2.40	1.14	4.66	0.51
22:6ω3	10.1	14.8	7.58	47.4

[a]DHASCO (DHA single cell oil) from Martek.

Table II. Omega-3 Fatty Acid Distribution in Different Positions of Triacylglycerols of Marine Oils.[a]

Fatty Acid	Menhaden Oil			Seal Blubber Oil		
	Sn-1	Sn-2	Sn-3	Sn-1	Sn-2	Sn-3
EPA	3.12	17.5	16.3	8.36	1.60	11.2
DPA	1.21	3.11	2.31	3.99	0.79	8.21
DHA	4.11	17.2	6.12	10.5	2.27	17.9

[a]Abbreviations are: EPA, eicosapentaenoic acid; DPA, docosapentaenoic acid; and DHA, docosahexaenoic acid.

Stability Issues

Oxidative stability and development of off-flavor in marine oils limit their widespread use in foods. Oxidation of highly unsaturated fatty acids in marine oils leads to the formation of an array of volatiles with very low threshold values. Thus, it is necessary to prevent their formation or to mask their undesirable odors in foods containing them. In this relation, use of antioxidants, encapsulation/ microencapsulation of the oil and use of novel packaging might provide the necessary means to extend the shelf-life of products.

To assess oxidative stability of edible oils in general and marine oils in particular, it is necessary to measure a) changes in the starting materials, b) production of primary products of oxidation and c) secondary oxidation products including individual volatile components, especially those with low threshold values (8). With respect to primary products of oxidation, a peroxide value of 5 and a Totox value (2PV + p-AnV) of 10 may be considered as the upper limit for oils intended for nutraceutical applications. However, omega-3 concentrates may readily reach these limits within a very short period of time, unless protected with a combination of appropriate storage conditions and use of effective antioxidants.

Changes in the Starting Material

Changes in the starting material may be monitored by measuring changes in the fatty acid composition or by weight gain or oxygen uptake, among others. While determination of composition of heavily oxidized marine oils which contain large proportions of highly unsaturated fatty acids is feasible, lightly oxidized oils which suffer from off-flavor defects may not exhibit much of a change in their fatty acid profile. However, measurement of oxygen uptake and weight gain or evaluation of induction period under accelerated conditions may provide useful information with regard to their oxidation potential (8).

Primary Products of Oxidation

Although determination of peroxide value (PV) is preferred, evaluation of lightly oxidized oils requires a few milliliters of sample. Thus, for colorless products, measurement of conjugated dienes which requires very little sample provides an excellent means for evaluation of their oxidative state on a comparative basis. However, for oils that are heavily colored, peroxide value determination provides the best means for their quality evaluation.

Nonetheless, it should be noted that primary products of oxidation do not have any color or off-flavor, despite their adverse health effects when present in the oils *(9)*.

Secondary Products of Oxidation and Individual Volatiles

The primary products of lipid oxidation, namely hydroperoxides, are labile, due to the low energy of their oxygen-oxygen bond, and degrade to an array of secondary products. These include hydrocarbons, aldehydes, ketones and alcohols, among others. Propanal is a dominant breakdown product of marine oils. Thus, use of propanal as an indicator for assessing the quality of marine oils is appropriate. However, evaluation of 2-thiobarbituric acid (TBA) reactive substances (TBARS), despite lack of specificity of this methodology, provides an excellent means for evaluation of the oxidation state of marine oils. Similarly, *para*-anisidine value (*p*-AnV) may be used for this purpose. Several authors have criticized the use of TBARS or *p*-AnV for evaluation of secondary products of oxidation *(10,11)*. However, in the opinion of this author, both of these indicators provide a valid means for evaluation of relative stability and quality of a specified product.

In terms of volatiles, produced from the breakdown of hydroperoxides, a wide array of products may be formed *(12)*. The most predominant aldehydes produced from oxidation of marine oils include (*Z*)-1,5-octadien-3-one (metallic), (*E,E,Z*)-2,4,7-decatrienal (fishy), (*Z*)-3-hexenal (green), and (*E,Z*)-2,6-nonadienal (cucumber), among others *(13)*. Of these, (*E,E,Z*)-2,4,7 decatrienal and its other isomers are predominant odorants in fish oils and seal oil. It is thought that presence of trimethylamine in oils from certain marine sources, particularly those from fish species containing trimethylamine oxide as an osmoregular, may be responsible for their fishy odor.

Stabilization of Marine Oils

Quality management of marine oils requires stabilization of products of interest. This may be achieved by a combination of proper storage and packaging of the materials involved along with the use of appropriate antioxidants. However, the original oils must be refined, bleached and deodorized to eliminate existing contaminants, including oxidation products. The potency of several antioxidants in stabilizing bulk marine oils is displayed in Table III. However, the industry often uses several antioxidants in combination and the efficacy of mixtures involved is indicated by the type of system (e.g. bulk, emulsion, low-moisture products) under consideration.

82

**Table III. Inhibition of Oxidation of Menhaden Oil by Selected
Antioxidants under Accelerated Storage Conditions as Reflected in
Thiobarbituric Acid Values.**

Antioxidant (ppm)	Oxidation Inhibition (%)
α-Tocopherol (500)	18.0
Mixed soy tocopherols (500)	32.5
TBHQ (200)	54.7
DGTE (200)	39.8
DGTE (1000)	48.5
EC (200)	45.1
EGC (200)	48.2
ECG (200)	51.3
EGCG (200)	50.0

[a]Abbreviations are: TBHQ, tert-butylhydroquinone; DGTE, dechlorophyllized green tea extract; EC, epicatechin; EGC, epigallocatechin; ECG, epicatechin gallate; and EGCG, epigallocatechin gallate.

In addition, trace metals, when present, may lead to rapid oxidation of the oils. Therefore, use of sequesting agents such as monoacylglycerol citrate may be considered. Furthermore, in certain systems, ternary systems involving tocopherol, ascorbic acid or ascorbyl palmitate and a phospholipid such as lecithin are most useful for stabilizing the products. However, highly unsaturated fatty acids may exhibit better stability when present in oil-in-water emulsions.

Masking of the off-odors in marine oils by other components in the formulations of foods, nutraceuticals and medicinal applications may be practical by use of encapsulation, microencapsulation and other proprietary techniques. The issue of rancidity in encapsulated health products, has nonetheless been the subject of recent research. Analysis of polyunsaturate-based nutraceuticals over the past ten years indicates a wide variation in quality. Therefore, use of antioxidants in such products is deemed necessary. The success of each technique depends on the nature of the specific formulation and the matrix involved. Possible presence of chlorophylls in crude plant-based extracts used for stabilization of oils might indeed exert a negative effect. Thus, crude green tea extracts behaved prooxidatively when applied to menhaden and seal blubber oils. Removal of chlorophyll circumvented this problem. Photosensitized oxidation when initiated by chlorophylls was found responsible for the former effect.

In functional food products, quality issues might be addressed by applying marine oils in foods that are usually consumed within a short period of time. Examples are bread and other baked products, milk and other dairy products as well as fabricated seafoods and above all infant formulas. Use of micro encapsulated oils in certain products might be necessary in order to extend the shelf-life of products.

Chitin, Chitosan, Chitosan Oligomers, *N*-Acetylglucosamine and Glucosamine

Chitin constitutes a major component of crustacean shells and accounts from up to 40% of the dry weight of the processing by-products of shrimp, crab, crayfish and lobster fisheries. Chitin is an analog of cellulose, but its monomer is 2-deoxy-2-acetamidoglucose (*14*). Following deproteinization (usually with 2-5% NaOH or KOH solution at moderate to high temperatures) and demineralization (usually with 2-5% HCl at room temperature), chitin is recovered (*15*). The quality of chitin is dictated by its ash and nitrogen contents as well as its color characteristics.

Chitosan, is the mainly deacetylated form of chitin (*16*). This product is prepared via the treatment of chitin with hot concentrated NaOH or KOH. The degree of deacetylation and the chain length of the polymer is affected by the temperature, concentration and duration of the process. The quality of chitosan is generally dictated by its molecular weight, as exhibited by its viscosity characteristics in solution. Chitosan, unlike chitin is readily soluble in weakly acidic solutions (*17*). For this purpose, acetic acid, lactic acid, citric acid and malic acid may be preferred as a means of solubilizing chitosan. Chitosan has a variety of health benefits and may be employed in a number of nutraceutical as well as other applications in the health industry (Table IV). The quality characteristics of chitosans depend on their molecular weight as reflected in their viscosity. This is due to the chain length of the molecules. The residual color, ash and/or nitrogen would also affect the quality of the final product.

Chitosan oligomers, prepared via enzymatic means, are readily soluble in water and hence may serve best in rendering their benefits under normal physiological conditions and in foods with neutral pH. Furthermore, depending on the type of enzyme employed, chitosan oligomers with specific chain length may be prepared to address the needs of the user industry.

Table IV.   Some Applications of Chitosans Materials.

Product	Effects
Chitosan	Hypocholesterolemic, weight loss, inhibitor of oxidation, nutrient and drug control release, etc.
Chitosan oligmers	Anti-inflammatory, antimicrobial, etc.
N-Acetylglucosamine	Anti-inflammatory, etc.
Glucosamine	Anti-inflammatory, etc.

The monomer of chitin, N-acetyl glucosamine (NAG) has been reported to possess antiinflammatory properties. Furthermore, the monomer of chitosan, glucosamine, is a commercial product marketed in the form of its sulfate salt.

Glucosamine sulfate has been reported to possess benefits for joint buildups. This product is prepared via hydrolysis of chitosan with hydrochloric acid. Products sold in the market as glucosamine sulfate, however, are often physical mixtures of glucosamine hydrochloride and sodium/potassium sulfate. However, the consumer is often unaware of this fact. The high salt content of such preparations (up to 30%) might be of concern. Recently, glucosamine was prepared via enzymatic means and this product is currently available in the Korean market (Kitolife Inc., Seoul, Korea). Furthermore, glucosamine products may also be sold in conjunction with chondroitins (chondroitin 4-sulfate and chondroitin-6-sulfate). Chondroitins are mucopolysaccharides (MPS) with molecular weight of up to 50,000 Da and could be prepared from connective tissues of slaughtered animals (18). In combination, while glucosamine helps to form proteoglycans that sit within the space in the cartiledge, chondroitin sulfate acts like a liquid magnet. Thus, glucosamine and chondroitin work together to improve the health of the joint cartiledge. Therefore, use of appropriate products and combinations is necessary when considering beneficial health effects of such preparations.

The by-products in the extract of chitin from shellfish are proteins, carotenoproteins and enzymes (19). These components may also be isolated for further value-added utilization. In particular, isolation and application of carotenoproteins, free xanthophylls and proteins is of interest.

Protein Hydrolyzates and Bioactive Peptides

Recovery of proteins from seafood processing discards may be achieved using alkali-assisted extraction procedures (*20*). However, it is also possible to extract these proteins following their hydrolysis. To achieve this, both chemical and enzymatic methods may be employed. However, due to the selectivity of enzymes, production of hydrolyzates via enzyme-assisted processes is preferred. Production of hydrolyzates using commercial enzymes is commonplace. However, depending on the degree of hydrolysis, certain products may exhibit bitterness. In such cases, inclusion of polar amino acids such as glutamic acid via plastein reaction would eliminate the bitterness of the products (*21*). Hydrolyzates may be added to food to improve their flavor characteristics (*22*) or act as alternative to polyphosphates in fabricated meat products (*23*). The amino acid sequence of certain peptides so produced may dictate their bioactivity (*24*). Although a similar sequence may exist in the protein, bioactivity of the peptide segments may not be exhibited in the original protein. Of particular interest is the antioxidant activity of peptides involved (*25*). These peptides not only serve as protectant of formulated foods against oxidation, but they also exert health benefits by boosting the immune system, among others. In fact, many of the antibiotics are short-chain peptides. Due to marked variability in the spectrum of possible products, quality characteristics of each product requires specialized procedures that go beyond the mandate of this cursory overview.

Algal and Microalgal Products

Marine algae and microalgae serve as an excellent source of alginate, protein, lipid, carotenoids as well as minerals such as potassium and iodine. Although seaweed has been used as a rich source of minerals and for extraction of alginates, recent efforts have been concentrated on the screening of varieties that can produce DHA in large enough quantities for commercialization. An example for this is DHA single cell oil (DHASCO) available from Martek with a DHA content of over 45%.

The use of algae and microalgae for therapeutic purposes has also a long history (*26*). The extracts of microalgal species have demonstrated antibacterial, antimicrobial, antiviral and antifungal activities. The active components were diverse and included fatty acids, bromophenols, peptides and polysaccharides (*27*). Numerous studies have also found health benefits for spirulina (e.g. *28*). These include cholesterol lowering, protection against nephrotoxicity, anticancer effect, radiation protection, stem-cell regeneration and immunomodulation effects. Quality management of algal species depends on the type of compound of interest. Thus, for DHA containing oils from algal sources, usual quality criteria for marine oils applies. However, for alginates and minerals, selection of criteria depends on the final application of these components.

Conclusions

Quality of marine nutraceuticals may be quite variable. As the consumers become more knowledgeable about characteristics of products, they demand high quality products. It is also of paramount importance to provide laboratory chemical analysis for standardization of products and to support efficacy of each nutraceutical by appropriate pre-clinical and clinical trials.

References

1. Wanasundara, U.N.; Shahidi, F. *J. Am. Oil Chem. Soc.* **1998**, *75*, 945-951.
2. Bazan, N.G.; Reddy, T.S.; Bazan, H.E.P.; Birkle, D.L. *Prog. Lipid Res.* **1986**, *25*, 595-606.
3. Newton, I.S. *J. Food Lipids* **1996**, *3*, 233-249.
4. Kyle, D.J. In *Omega-3 Fatty Acids: Chemistry, Nutrition and Health Effects*; Shahidi, F.; Finley, J.W., Eds.; ACS Symposium Series 788, American Chemical Society, Washington, DC., 2001, 92-107.
5. Kanayasu-Toyoda, T.; Morita, I., Murota, S.I. *Prostaglandins, Leukotrienes and Essential Fatty Acids* **1996**, *54*, 319-325.
6. Wanasundara, U.N.; Shahidi, F. *Food Chem.* **1999**, *65*, 41-49.
7. Shahidi, F.; Wanasundara, U.N. *Trends Food Sci. Technol.* **1998**, *9*, 230-240.
8. Shahidi, F.; Wanasundara, U.N. In *Food Lipids: Chemistry, Nutrition and Biotechnology*; Akoh, C.C.; Min, D.B., Eds.; Marcel Dekker Inc.; New York, 1998b; pp. 377-396.
9. Shahidi, F.; Wanasundara, U.N. *Food Sci. Technol. Int.* **1996**, *2*, 73-81.
10. Dugan, L.R. *J. Am. Oil Chem. Soc.* **1955**, *32*, 605-609.
11. Baumgartner, W.A.; Baker, N.; Hill, V.A.; Wright, E.T. *Lipids* **1975**, *10*, 309-311.
12. Senanayake, S.P.J.N.; Shahidi, F. In *Omega-3 Fatty Acids: Chemistry, Nutrition and Health Effects*; Shahidi, F.; Finley, J.W.; Eds.; ACS Symposium Series 788; American Chemical Society: Washington, DC; 2001, pp. 151-173.

13. Cadwallader, K.R.; Shahidi, F. In *Omega-3 Fatty Acids: Chemistry, Nutrition and Health Effects*; Shahidi, F.; Finley, J.W.; Eds.; ACS Symposium Series 788, American Chemical Society: Washington, DC, 2001, pp. 221-234.
14. Lower, E.S. *Manufac. Chem.* **1984**, *55*, 73-75.
15. Shahidi, F.; Synowiecki, J. *J. Agric. Food Chem.* **1991**, *39*, 1527-1532.
16. Lower, E.S. *Manufac. Chem.* **1984**, *55*, 47-82.
17. Bough, W.A. *Food Prod. Dev.* **1977**, *11*, 90-92.
18. Synowiecki, J.; Shahidi, F. *Food Chem.* **1994**, *51*, 89-93.
19. Simpson, B.; Nguyen, T.T.D.; Awafo, V. In *Seafood in Health and Nutrition*; Shahidi, F.; Ed.; Science Tech Publishing Company: St. John's, Canada, 2000, pp. 195-203.
20. Shahidi, F.; Synowiecki, J. *Food Chem.* **1996**, *57*, 317-321.
21. Synowiecki, J.; Jegielka, R.; Shahidi, F. *Food Chem.* **1996**, *57*, 435-439.
22. Hwang, C-F.; Shahidi, F.; Onodenalore, A.C.; Ho, C-T. In *Flavor and Lipid Chemistry of Seafoods.* Shahidi, F.; Cadwallader, K.R., Eds. ACS Symposium Series 674; American Chemical Society: Washington, DC, 1997, pp. 76-84.
23. Shahidi, F.; Synowiecki, J. *Food Chem.* **1997**, *60*, 29-32.
24. Kim, S-K.; Tang, Y-T.; Kim, Y-T.; Byun, H-G.; Nam, K-S.; Joo, D-S.; Shahidi, F. *J. Agric. Food Chem.* **2001**, *49*, 1984-1989.
25. Amarowicz, R.; Shahidi, F. *Food Chem.* **1997**, *58*, 355-359
26. Hoppe, H.A. In *Marine Algae in Pharmaceutical Science.* Hoppe, H.A.; Lerring, T.; Tanaka, Y. Eds. Walter de Gruyter, New York, 1979, 25-119.
27. Borowitzka, M.A. *J. Appl. Physiol.* **1995**, *7*, 3-15.
28. Belag, A.; Kato; T.; Ota, Y. *J. Appl. Physiol.* **1996**, *8*, 303-311.

Analysis of a Specific
Nutraceutical

Chapter 7

Species Identification of Black Cohosh by LC–MS for Quality Control

Kan He, Bo Lin Zheng, Calvin Hyungchan Kim, Ling-ling Rogers, and Qun Yi Zheng*

Department of Research and Development, Pure World Botanicals, Inc., 375 Huyler Street, South Hackensack, NJ 07606

A method to directly identify triterpene glycosides using reversed-phase liquid chromatography with positive atmospheric pressure chemical ionization mass spectrometry (LC/(+)APCIMS) was developed. Based on the analysis of the molecular weight, fragment ions, and selected ion chromatograms, a number of triterpene glycosides, including actein (1), 27-deoxyactein (2), cimicifugoside M (3), and cimicifugoside (4), from *Cimicifuga racemosa* were studied. Cimicifugoside M was further isolated and was identified as a new triterpene glycoside, cimigenol 3-*O*-α-L-arabinopyranoside. Meanwhile, a chromone, cimifugin (5), was also isolated from *C. foetida*. Cimicifugoside M and cimifugin can serve as markers for species identification. The method can therefore be used to distinguish black cohosh products from different Cimicifuga species for quality control purposes. The four triterpene glycosides (1-4) were evaluated for the estrogen-receptor-binding activity and it was found that they do not bind to estrogenic receptors.

Introduction

Cimicifuga racemosa (L.) Nutt (Ranunculaceae), commonly known as black cohosh, is a Native American medicinal plant. Its roots and rhizomes have long been used to treat a variety of women's complaints, especially painful periods, menopausal disorders, premenstrual complaints, dysmenorrhea and uterine spasms in the 19th century. The modern preparation of black cohosh extract is the major component found in commercial European preparations for menopausal disorder and has been widely used in Europe for over 40 years in over 1.5 million cases (*1*). Numerous clinical studies using Remifemin for the black cohosh extract have confirmed its efficacy as an alternative to estrogen treatment. According to the German Commission E Monographs, black cohosh consists of dried rhizomes and roots of *Cimicifuga racemosa* (*2*). Currently, the most common commercially available preparation of black cohosh is a powdered extract which is standardized for its content of triterpene glycosides (calculated as 27-deoxyactein). Some products using Asian *Cimicifuga* species, such as *C. foetida*, *C. simplex*, or a combination of species are also found on the marketplace. An HPLC analytical method with UV and ELSD detectors for analysis of triterpene glycosides has been developed. However, this system is still unable to identify triterpene glycosides using the above mentioned detectors without the availability of standards of each peak. This leads to difficulty of species identification due to the complicated HPLC chromatograms they presented. Therefore, it is necessary to find some unique marker compounds which can serve as specific indicators to distinguish different species in the genus. Not much chemical information on *Cimicifuga racemosa* can be found in early literature. Only actein (**1**), 27-deoxyactein (**2**), and cimicifugoside (**4**) were reported from black cohosh (*3-7*). We have isolated these three major triterpene glycosides during the analytical method development (Figure 1). In addition, the compound cimicifugoside M (**3**), was isolated and identified as a new triterpene glycoside based on NMR spectroscopic data. A number of triterpene structures have been recently elucidated by two groups (*8-9*), adding to 18, the total number of triterpene glycosides in the black cohosh.

The presence of complicated matrix interferences in the black cohosh extract and low UV absorption due to the absence of a strong chromophore in the triterpene structures makes the identification of these compounds difficult. Using an atmospheric pressure chemical ionization (APCI) as an HPLC detector has become the method of choice for the analysis of triterpene glycosides in black cohosh. APCI is an ionization technique in which the analyzed molecules react with reagent ions at atmospheric pressure. The reagent ions are formed via solvent vaporization and subsequent interaction with high voltage, these creating a corona discharge. The formed ions are then transmitted through an electrostatic field and captured in the ion trap. APCI is a soft ionization technique which provides little or no fragmentation for thermally stable compounds. However, due to the high temperature needed for vaporization (450°C), both the molecular ion and fragmentation ions of the thermally unstable triterpene glycosides will be detected by mass spectrometer. *Cimicifuga* species are known to contain highly oxygenated triterpene glycosides. The oxygen atoms provide the important fragment ions which are derived from the loss of a water, a sugar, or an acetyl group. Identification of the compound is based on a group of fragment ions. We have used this method to identify several triterpene glycosides and found that one triterpene, cimicifugoside M, which was only detected in *C. racemosa*, and a non-triterpene glycoside, cimifugin, which was only detected in *C. foetida* and *C. simplex* can specifically serve as markers for species differentiation. We have applied this method to check more than 30 commercially available black cohosh powder extracts. The results of our analyses are presented and discussed in the following sections.

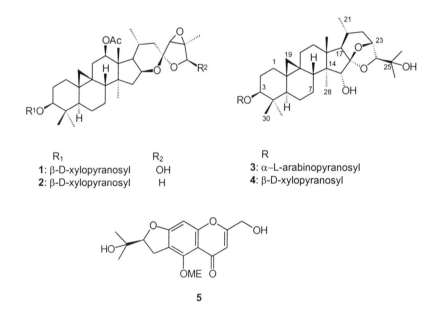

	R_1	R_2
1:	β-D-xylopyranosyl	OH
2:	β-D-xylopyranosyl	H

	R
3:	α–L-arabinopyranosyl
4:	β-D-xylopyranosyl

5

Figure 1. Triterpene glycosides (1-4) isolated from C. racemosa. Cimifugin (5) isolated from C. foetida.

Identification of New Triterpene Glycoside

The four pure triterpene glycoside standards used in the LC/MS experiment were actein (**1**), 27-deoxyactein (**2**), cimicifugoside M (**3**) and cimicifugoside (**4**). These standards were isolated from *Cimicifuga racemosa* and were compared by ^{1}H and ^{13}C NMR data with literature (*10-13*). Cimicifugoside M is a new compound which was first isolated from *C. racemosa* and the spectral data are listed as follows. HR-FAB-MS: found: 621.3996; calcd. for $C_{35}H_{56}O_9$ 621.3927. IR ν_{max} 3448, 2942, 1642, 1455, 1383, 1255, 1167, 1149, 1070 cm^{-1}. ^{1}H-NMR (400 MHz, pyridine-d$_5$): δ = 1.24, 1.57 (m, 2H, H-1), 1.94 (qd, 1H, J = 11.6, 3.6 Hz, H-2), 2.36 (m, 1H, H-2), 3.48 (dd, 1H, J = 11.6, 4.4 Hz, H-3), 1.31 (dd, 1H, J = 12.4, 3.6 Hz, H-5), 0.71 (br. q, 1H, J =12.4 Hz, H-6), 1.51 (m, 1H, H-6), 1.19, 2.06 (m, 2H, H-7), 1.70 (m, 1H, H-8), 1.08, 2.05 (m, 2H, H-11), 1.54, 1.68 (m, 2H, H-12), 4.24 (d, 1H, J = 8.6 Hz, H-15), 1.50 (m, 1H, H-17), 1.14 (s, 3H, H-18), 0.27, 0.51 (d, 2H, J = 4.0 Hz, H-19), 1.68 (m, 1H, H-20), 0.84 (d, 3H, J = 6.4 Hz, H-21), 1.04, 2.25 (m, 2H, H-22), 4.74 (d, 1H, J = 9.2 Hz, H-23), 3.75 (s, 1H, H-24), 1.45 (s, 3H, H-26), 1.47 (s, 3H, H-27), 1.17 (s, 3H, H-28), 1.26 (s, 3H, H-29), 1.01 (s, 3H, H-30), 4.78 (d, 1H, J = 7.0 Hz, H-1'), 4.44 (dd, 1H, J =7.4, 7.0 Hz, H-2'), 4.15 (br. d, 1H, J = 7.6 Hz, H-3'), 4.31 (br. s, 1H, H-4'), 3.78, 4.29 (dd, 2H, J = 10.4 Hz, H-5'). ^{13}C NMR (400 MHz, pyridine-d$_5$): δ = 32.46 (C-1), 30.11 (C-2), 88.63 (C-3), 41.38 (C-4), 47.63 (C-5), 21.10 (C-6), 26.38 (C-7), 48.66 (C-8), 20.04 (C-9), 26.70 (C-10), 26.48 (C-11), 34.11 (C-12), 41.88 (C-13), 47.32 (C-14), 80.25 (C-15), 112.00 (C-16), 59.61 (C-17), 19.54 (C-18), 30.90 (C-19), 24.13 (C-20), 19.61 (C-21), 38.19 (C-22), 71.86 (C-23), 90.21 (C-24), 70.98 (C-25), 25.44 (C-26), 25.21 (C-27), 11.84 (C-28), 27.77 (C-29), 15.43 (C-30), 107.49 (C-1'), 72.98 (C-2'), 74.68 (C-3'), 69.55 (C-4'), 66.77 (C-5').

All the protons and protonated carbons in **3** are assigned based on a combination of data obtained by DEPT, ^{1}H-^{1}H COSY, TOCSY, HMQC, HSQC, and ROSEY NMR spectra. These assignments are supported by the close similarity of the ^{1}H and ^{13}C NMR spectroscopic data with those obtained for cimicifugoside (**4**). The partial structures deduced from different spin systems by the ^{1}H-^{1}H correlation spectroscopy are connected by analysis of an HMBC spectrum. As shown in Figure 2, the methyl protons at δ 1.26 and 1.01 (CH$_3$-29 and CH$_3$-30) show long-range correlations with the carbons at δ 41.38 (C-4), 88.63 (C-3), and 47.63 (C-5). This indicates that C-4 is connected to C-29, C-30, C-3, and C-5. The methyl protons at δ 1.14 (CH$_3$-18) exhibit long-range correlations with the carbons at δ 41.88 (C-13), 34.11 (C-12), 47.32 (C-14), and 59.61 (C-17), suggesting that C-18, C-12, C-14, and C-17 are connected with C-13. Similarly, the correlation of the protons at CH$_3$-28 (δ 1.17) with C-14, C-13, and C-15 (δ 80.25) indicates that C-14 is connected to C-28, C-13, and C-15. Also, the methyl signal at δ 0.84 (CH$_3$-21) is correlated with the signals at δ 24.13 (C-20), 59.61 (C-17), and 38.19 (C-22), indicating that C-20 is connected to C-21, C-17, and C-22. Furthermore, methyl signals at δ 1.45 (CH$_3$-26) and 1.47 (CH$_3$-27) display long-range correlations with carbons at δ 70.98 (C-25) and 90.21

(C-24) indicating that C-25 is connected with C-26, C-27,and C-24. These correlations give further evidence that **3** possesses a cimigenol aglycone.

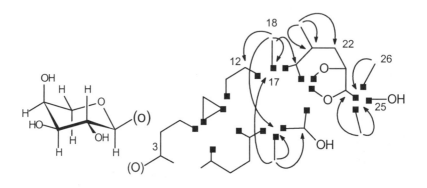

*Figure 2. Partial structures of **3** and significant long-range correlations observed in the HMBC spectrum.*

The relative stereochemistry of **3** is determined by analysis of the ROSEY and NOE difference spectra as well as the coupling constants. All the NOE correlations indicate the structural configuration as shown in the Figure 3. Combining all the spectroscopic data, cimicifugoside M is determined to be cimigenol 3-*O*-α-L-arabinopyranoside.

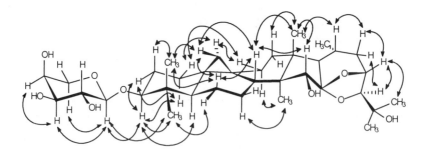

*Figure 3. Structural configuration of **3** and some important NOE correlations observed in the ROESY and NOE difference spectra.*

LC/MS Analysis of the Triterpene Glycosides and Cimifugin

Structurally, there are two major types of triterpene glycosides based on the literature report. Actein with acteol aglycone contains a xylosyl and cimicifugoside with cimigenol aglycone which contains a xylosyl or arabinosyl. LC/(+) APCI-MS experiments were initially carried out on four pure triterpene glycosides (**1-4**). Both actein and 27-deoxyactein contain an acetyl group at the 12-position; therefore, the loss of an acetyl group provides useful diagnostic information for the presence of these two compounds. In the mass spectrum of actein (Figure 4), the pseudo molecular ion $[MH]^+$ was observed at m/z 677. The other major peaks observed at m/z 659 and 599 were derived from the loss of water (18 amu) from $[MH]^+$ and the subsequent loss of acetic acid (60 amu), respectively. A peak at m/z 527 is derived from a loss of xylosyl from $[MH]^+$, which provides the evidence of the existence of acteol aglycone. Other fragment ions observed at m/z 467, 449 and 432 were formed from $[MH]^+$ by loss of an acetyl and xylosyl group followed by two successive losses of water. Some minor peaks, due to a loss of water or acetic acid from $[MH]^+$ or other fragment ions, were also observed. The suggested pathway involves the protonation of an oxygen atom followed by a series of neutral molecule losses. The oxygen atom in the ring undergoes ring opening followed by water loss. In the spectrum of 27-deoxyactein, the major peak which is due to a loss of an acetyl group was observed at m/z 601. The loss of an additional xylosyl group followed by three successive losses of water results in the fragments observed at m/z 469, 451, 433 and 415, respectively. Since the difference in molecular weight between actein and 27-deoxyactein is 16 amu, the loss of a water molecule (-18 amu) always produces a fragment with a m/z difference of 2 amu when compared to the corresponding fragment ions. This provides diagnostic information which is useful in distinguishing these two compounds. In the mass spectra of cimicifugoside M and cimicifugoside, the pseudo molecular ion $[MH]^+$ is observed at m/z 621. Loss of water molecules from both the molecular ions and the fragment ions gives the characteristic fragmentation patterns of these two compounds. The structural difference between cimicifugoside M and cimicifugoside is due to the different sugar moieties, arabinosyl and xylosyl. Since the molecular weights of the sugars are the same, it is not possible to distinguish them based on the fragment ions, but they can be distinguished by the differences in retention time.

The protonated molecular ion of actein $[MH]^+$ observed at m/z 677 is weak, but an intense series of peaks is produced by a group of characteristic fragment ions generated by neutral molecule losses. These fragment ions which are used for identification purposes are observed at m/z 677, 659, 617, 599, 581, 527, 467, 449 and 431. Similarly, identification of 27-deoxyactein can be made based on the presence of the molecular ion and other fragment ions observed at m/z 661, 601, 583, 469, 451, 433 and 415. Peaks observed at m/z 621, 603, 585, 512, 489, 471, 453, 435 and 417 indicated the presence of cimicifugoside M or cimicifugoside.

An alcoholic extract of *C. racemosa* was scanned under the (+)APCIMS mode and more than 20 peaks were resolved by the HPLC condition. Many of them could be positional, steric isomers, or structural difference, such as xylosyl and arabinosyl substitute, which have the same molecular weight and similar fragment patterns to those of actein, 27-deoxyactein, or cimicifugoside. As discussed previously, these fragments were derived from the water, sugar, or acetyl loses and only those peaks which showed the same fragment patterns with standards were related to triterpene glycosides. In this way, we found that all of the major peaks seen in the total ion current, displayed black cohosh triterpene fragment patterns. Using the molecular ion and fragment ion information obtained from (+) APCI/MS, most triterpene glycosides of *C. racemosa* can be classified. However, the absolute structural identification of each peak from LC still has to be based on the direct comparison of each standard.

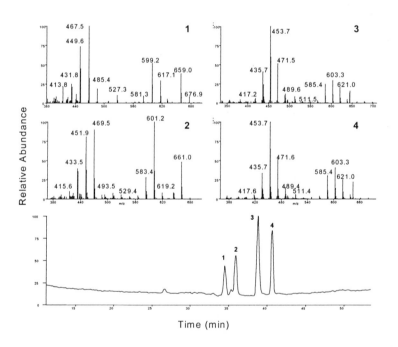

Time (min)

Figure 4. Total ion chromatogram and mass spectra of actein (1), 27-deoxyactein (2), cimicifugoside M (3), and cimicifugòside (4).

Two compounds were eluted (t_R around 12 min and 13 min) in the alcoholic extract of *C. foetida*. The UV spectra of both compounds exhibits absorption maxima at the 233, 251, 256 and 298 nm. These two components were not related to triterpene glycosides since no characteristic fragments were found. The first peak shows the MH ion at m/z 469 and its aglycone MH at m/z 307 derived from losing a glucosyl while the second peak (**5**) only displays the molecular ion at m/z 307 in the mass spectra. We isolated **5** from *C. foetida* and its structure was elucidated as cimifugin based on the spectroscopic method, mainly NMR. Both these two compounds, cimifugin and cimifugin 7-*O*-β-D-glucoside, were previously reported to be isolated from *C. simplex* and *Angelica japonica* (*14-15*). These two peaks were only detected in *C. foetida* and *C. simplex* but not in *C. racemosa* by LC/MS.

Species Identification by LC/MS

Some Asian *Cimicifuga* species have been used in traditional Chinese medicine and Japanese folk medicine as antipyretic, analgesic, and anti-inflammatory agents. More than fifty triterpene glycosides were reported from these species. Due to the complexity of component interference, it is not sufficient to analyze these components just based on chromatographic retention time and molecular weight. Thus, a direct comparison of HPLC chromatograms can not provide diagnostic information for species differentiation (Figure 5).

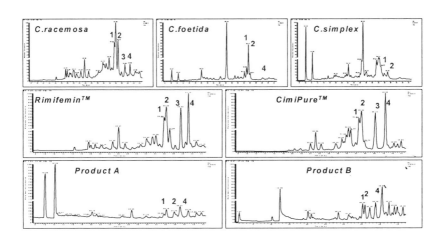

Figure 5. Comparison of total ion chromatograms of different Cimicifuga species and black cohosh products.

With the introduction of the characteristic fragment ion patterns into the analysis, the peak determination will be less ambiguous when peak co-elution and HPLC retention time shifts happen. Therefore, a selected ion chromatogram (SIC), which was obtained based on the four standards (1-4) is used for species identification. Four peaks which correspond to 1-4 at m/z 599 (m/z 599 is chosen due to its high intensity), m/z 661, and m/z 621 were detected in SIC of *C. racemosa*. 1, 2 and 4 were also found in the alcoholic extract of *C. foetida* and *C. simplex* while 3 was not detected in a noticeable concentration. It is likely that 3 can serve specifically as an marker for identifying *C. racemosa* species. Similarly, 5 can be a marker for *C. foetida* and *C. simplex*. We have tested more than 30 commercially available black cohosh products with this technique and three types of products have been found. They are products using *C. racemosa* seen in Remifemin and CimiPure where 3 was detected and 5 was not (Figure 6) and products using *C. foetida* or *C. simplex* where only 5 was detected and 3 was not detected (products A and B). The third type of the product was a combination of *C. racemosa* and *C. foetida* or *C. simplex* in which both 3 and 5 were found.

Discussion

Currently, the black cohosh extract is mainly used for premenstrual syndrome and menopause associated with hot flashes, uterine spasms, depression, heart palpitation, night sweat, insomnia, and dysmenorrhea, etc. As a woman approaches menopause, the signals between the ovaries and pituitary gland diminish. Declining levels of estrogen stimulates the release of luteinizing hormone (LH) through the feedback mechanisms. A sudden burst of LH will cause the above mentioned symptoms. Usually, these symptoms are treated by estrogen replacement therapies (ERT). Since more and more studies relate that ERT has a high risk of breast cancer, many women are unwilling to take estrogens in high dosage and for a long duration. Thus, as an alternative choice, black cohosh becomes important for menopausal women. The 1989 German Commission E Monographs has indicated that the actions of black cohosh include estrogen-like activity, LH suppression, and binding to estrogen receptors which were based on early pharmacological studies (*1*). Due to the similarity in structures of estrodiol or synthetic estrogens such as diethylstilbestrol, the triterpene glycosides and isoflavonoids in black cohosh have been speculated as active components with estrogen mimetic activity.

In our study, however, no binding was observed for the pure actein, 27-deoxyactein, cimicifugoside M, and cimicifugoside in the estrogen receptor binding assay. This result indicates the possibilities that: 1) triterpene glycosides are not active components in the black cohosh extract; 2) triterpene glycosides are active but with different mechanism; or 3) a synergistic action is implicated with other components. So far, the pharmacological properties of triterpene glycosides in general are not clear. Therefore, it may not be appropriate to extrapolate the clinical research which was obtain from *Cimicifuga racemosa* to other *Cimicifuga* species since the chemical composition and clinical effectiveness may not be the same. In addition, the early studies had reported that an isoflavonoid, formononetin, responds to the estrogenic activity in black cohosh (*16*). But we can not detect the existence of formononetin in an appreciable concentration in the black cohosh extract. Formononetin is a pro-estrogen, being converted *in vivo* to the more active isoflavan, equol. Equol has weak estrogenic propriety. If formononetin responses to the activity, the concentration to achieve a biological activity would have to be relatively high to be detected. Contrary to the early report, the latest studies have indicated that black cohosh does not have estrogen-like action and does not suppress LH releasing (*1*).

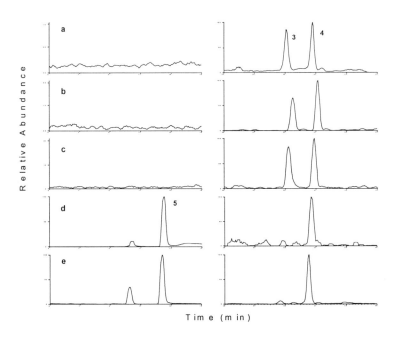

Figure 6. Comparison of selected ion mode at m/z 307 for cimifugin (5), m/z 621 for cimicifugoside M (3) and cimicfugoside (4). a) Cimicifuga racemosa. b) Remifemin. c) CimiPure. d) Cimicifuga foetida. e) Product A.

Although the mechanism of action is still unclear, *Cimicifuga racemosa* has been demonstrated by a number of clinical trials in the treatment of premenstrual and menopausal syndromes. Therefore, in order to make claims based on the European clinical trials for a product with black cohosh, the right species should be used to achieve the efficacy and safety.

The method demonstrated here should have widespread applications in the analysis of natural products, especially in the identification of triterpene saponins that have low absorbance in the UV wavelength range. It is our goal to develop a quick and accurate method for the analysis of triterpene glycosides in black cohosh for quality control.

References

1. Foster, S. *Herbal Gram* **1999**; *45*, 36-49.
2. Blumenthal, M.; Busse, W. R.; Goldberg, A.; Gruenwald, J.; Hall, T.; Klein, S.; Riggins, W.; Rister, R. S., Eds.; *The Complete German Commission E Monographs, Therapeutic Guide to Herbal Medicines, American Botanical Council*, Austin, Texas, Integrative Medicine Communications, Boston, Massachusetts, 1998; p90.
3. Corsano, S.; Piancatelli, G.; Panizzi, L. *Gazz. Chim. Ital.* **1969**, *99*, 915-932.
4. Radics, L.; Kajtar-Peredy, M.; Corsano, S.; Standoli, L. *Tetrahedron Lett.* **1975**, *48*, 4287-4290.
5. Piancatelli, G. *Gazz. Chim. Ital.* **1971**, *101*, 139-48.
6. Corsano, S.; Mellor, J. M.; Ourisson, G. *Chem. Comm.* **1965**; (10) 185-186.
7. Linde, H. *Arch. Pharmaz.* **1968**, *301*, 335-341.
8. Bedir, E.; Khan, A. I. *Chem. Pharm. Bull.* **2000**, *48*, 425-427.
9. Shao, Y.; Harris, A.; Wang, M.; Zhang, H.; Cordell, G. A.; Bowman, M.; Lemmo, E. *J. Nat. Prod.* **2000**, *63*, 905-910.
10. Koeda, M.; Aoki, Y.; Sakurai, N.; Nagai, M. *Chem. Pharm. Bull.* **1995**, *43*, 1167-1170.
11. Kusano, A.; Shibano, M.; Kitagawa, S.; Kusano, G.; Nozoe, S.; Fushiya, S. *Chem. Pharm. Bull.* **1994**, *42*, 1940-1943.
12. Kusano, A., Shibano, M., Kusano, G.; *Chem. Pharm. Bull.* **1996**; *44*, 167-172.
13. Kadota, S.; Li, J-X.; Tanaka, K.; Namba, T. *Tetrahedron* **1995**, *51*, 1143-1146.
14. Kondo, Y.; Takemoto, T. *Chem. Pharm. Bull.* **1972**, *20*, 1940-1944.
15. Baba, K.; Hata, K.; Kimura, Y.; Matsuyama, Y. *Chem. Pharm. Bull.* **1981**, *29*, 2565-2570.
16. Jarry, H.; Harnischfeger, G.; Düker, E. *Planta Medica* **1985**, *51*, 316-319.

Chapter 8

Modern Analytical Approaches in Quality Control of Black Cohosh

R. Spreemann[1], H. Kurth[2], M. Unger[2], and F. Gaedcke[2]

[1]Finzelberg, Inc., 2 Sylvan Way, Parsippany, NJ 07054
[2]Finzelberg GmbH and Company KG, 56626 Andernach, Germany

Quality and reproduction of the constituents of an herbal drug preparation are of great importance to assure efficacy and safety of the resulting herbal medicinal product. Therefore, it is necessary to have optimized and validated analytical methods. For quality control the phytochemical characterization of black cohosh rhizome and extracts focuses on the identification and quantitative determination of triterpene glycosides. Quantitative determination of the important triterpene glycosides actein, 27-desoxyactein, cimicifugoside and cimicifugoside M is done by RP-HPLC with Evaporative Light Scattering Detection (ELSD). Further marker substances like isoferulic acid are quantified if necessary by HPLC.

Black cohosh was introduced to Americans by Native Americans calling it "squaw root" or "snake root". Because it promotes and restores healthy menstrual activity it was predominantly used in treating uterine disorders and other female complaints (*1*).

In 1989 the German Commission E published the *Cimicifuga racemosa* monograph indicating the use of black cohosh for treating premenstrual, dysmenorrhic and menopausal vegetative symptoms (*2*). The recommended daily dosage of Commission E is 40 mg herbal drug or equivalent herbal drug preparations being extracted with ethanol water 40–60%. Today the use of black cohosh focuses on the treatment of menopausal complaints and PMS (*3*).

It is already very popular in Europe and we expect to see increased usage in the United States. Black cohosh could become of increasing interest to women looking for an alternative to estrogen therapy for the treatment of menopausal symptoms in the near future. Therefore it is very important to establish powerful analytical tools to describe the quality of the extract and hence to guarantee efficacy.

Constituents in Black Cohosh

Black cohosh contains several important constituents (Figure 1). Among them, triterpene glycosides actein and deoxyactein, as well as cimicifugoside and cimicifugoside M are considered the main pharmacologically relevant constituents. Actein differs from deoxyactein in that it contains an additional hydroxyl group in position 27. The cimicifugosides contain a cyclopropane ring as a common structural feature, and are structurally related to cycloartenol. The structural difference between cimicifugoside M and cimicifugoside is due to the different sugar moieties (*4*).

In a recent remarkable study a systematical investigation of triterpene glycosides was published (*5*). The authors reported about the isolation and structure of sixteen triterpene glycosides. Eight of them were described for the first time.

Further *Cimicifuga racemosa* contains cinnamic acids, such as caffeic, ferulic and isoferulic acid. In the past there have been controversial reports by several authors about the isoflavone formononetin as a major active constituent assuming to have estrogen-like effects (*6-8*). According to our investigation formononetin occurs in the root in such small quantities, that it is hardly detectable in extracts with conventional HPLC methods.

Further constituents are tannins, resins, fatty acids, starch and polysaccharides (*9*).

Until today it was not possible to assign the active principle definitely to one or more of the constituents described above. However there are pharmacological hints that show that the triterpene glycosides might participate in the efficacy.

Monographs of Black Cohosh

As shown in Table I there are only a few official monographs available for black cohosh rhizomes and extracts. The British Herbal Pharmacopoeia 1996 (BHP 1996), the German Homeopathic Pharmacopoeia (HAB 1) and the French Pharmacopoeia (for homeopathic preparations 1989) describe the herbal drug.

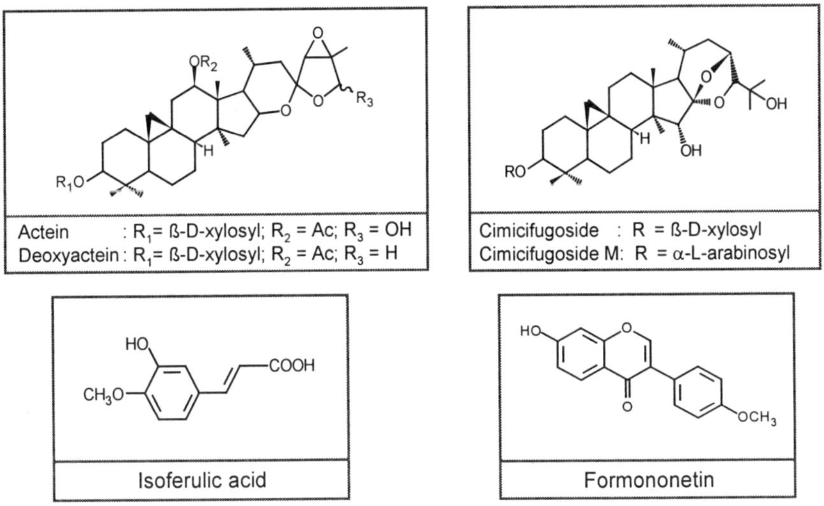

Actein : R_1= ß-D-xylosyl; R_2 = Ac; R_3 = OH Deoxyactein: R_1= ß-D-xylosyl; R_2 = Ac; R_3 = H	Cimicifugoside : R = ß-D-xylosyl Cimicifugoside M: R = α-L-arabinosyl
Isoferulic acid	Formononetin

Figure 1. Structures of several important constituents of black cohosh.

Usually only the description of the rhizomes (macroscopically and microscopically) is given as well as some additional parameters for purity (total ash, insoluble ash, loss on drying). The identity is checked by thin layer chromatography (TLC), but there are no requirements for minimum contents of triterpene glycosides or other constituents in these monographs

The WHO draft monograph published in 1998, covers in addition to the common parameters mentioned testing on pesticides, heavy metals and radioactivity. A quantification of constituents was not required. This is because the aforementioned constituents were not thought to contribute to the efficacy of black cohosh until now. Therefore, the total extract, in its entirety is considered as the active substance.

Dosage recommendations are given only by the Commission E monograph and the WHO draft monograph (1998). WHO has adopted the Commission E recommendations.

The daily dosage for the extract is equivalent to 40 mg herbal drug following Commission E and WHO.

Table I. Official Monographs of Black Cohosh

Monograph	Description		Identification	Foreign Matter [%]	Ash [%]	AIA [%]	LOD [%]	Pesticides	Heavy metals	Assay	Recommended daily dosages
	macro	micro	TLC								
BHP 96	+	+	+	≤2 ≤5	≤10	≤4	-	-	-	-	-
HAB 1	+	-	-	-	-	-	-	-	-	-	-
HPUS 90										-	
PF 10 (homéopatique)	+	+	+	-	≤6	-	-	-	-	-	-
ESCOP	-	-	-	-	-	-	-	-	-	-	-
Commission E	(+)	-	-	-	-	-	-	-	-	-	Extracts corresponding to 40 mg of drug
WHO-draft 10/23/98	+	+	-	≤2 ≤5	≤10	≤4	≤12	Ph.Eur.97 Aldrin + Dieldrin ≤0.05 ppm	Pb:10 Cd: 0.3	-	Extracts corresponding to 40 mg of drug
Finzelberg	+ Erg.B.6	+ Erg.B.6	+	≤5	≤10	-	≤12	Ph.Eur. 97	Pb: 5 Cd: 0.3 – 0.5 Hg: 0.1 ppm	+	-

Quality Control Aspects of Black Cohosh

In Europe quality control of herbal medicinal products is based on national and international regulations. Guidelines such as "Notice to Applicants" and "Quality of Herbal Medicinal Products" describe all relevant aspects of quality. Therefore, the herbal drug, and consequently the resulting preparation, have to be tested according to detailed specifications in order to guarantee safe and efficacious products. The following chapter describes some aspects of modern quality control of black cohosh extracts.

Identity

The TLC fingerprint of black cohosh constituents used is shown in Figure 2. The method used is based on BHP 1996. Slightly above the zone of rutin in the test solution is a prominent blue fluorescent zone. Slightly below the zone of isoferulic acid a prominent blue fluorescent zone is also visible.

Figure 3 shows a TLC fingerprint of triterpene glycosides, developed by Finzelberg. In the upper third of the chromatogram of the test solution are blue fluorescent zones visible at the level of the zone of isoferulic acid. At the Rf value of approximate 0.3 a strong blue fluorescent zone occurs. The number of bands and hence the value of information is much better than in the example before.

However, both TLC-methods are very suitable for identification. Additionally they were proved successful in monitoring the fingerprint in stability tests. But due to the complex nature of the triterpene pattern in black cohosh it is very difficult to make sound judgments about possible admixtures with other species like *Cimicifuga dahurica*, *C. foetida*, *C. simplex* etc.

Assay

Marker compounds are valuable tools for definition purposes in quality control. Possible uses are identity and purity tests, batch specific control, determination of transition rate, validation of manufacturing conditions, stability tests and proof of batch conformity (*10*). The caffeic acid derivative isoferulic acid is not as ubiquitous in nature as the isomeric ferulic acid. As it is also commercially available in sufficient quantities isoferulic acid is an ideal

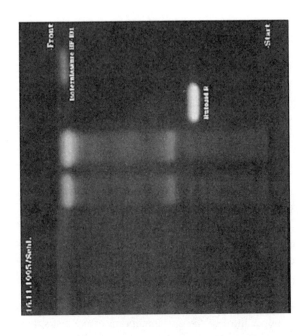

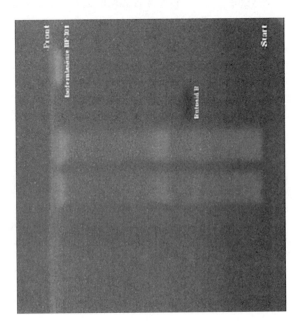

Figure 2. TLC fingerprint of black cohosh extract acc. to British Herbal Pharmacopoeia (Left: UV 365 nm, right: UV 365 nm after derivatization).

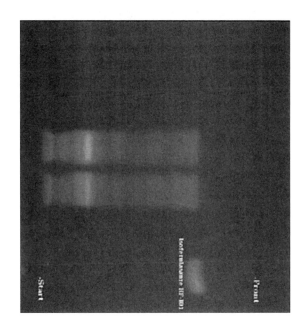

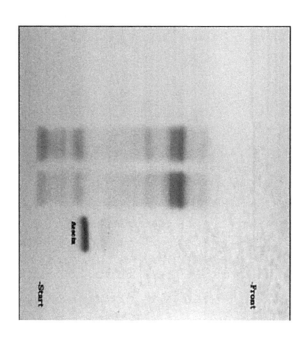

Figure 3.: TLC fingerprint of black cohosh extract according to Finzelberg (Left: UV 365 nm, right: Daylight after derivatization).

compound to serve as a marker in batch specific control. Figure 4 shows a chromatogram of the separation of caffeic acids occurring in black cohosh as an example for batch specific control. The chromatogram can be achieved with a simple acetonitril-water-phosphoric acid mixture on a RP18 column. Isoferulic acid can be easily detected at 319 nm.

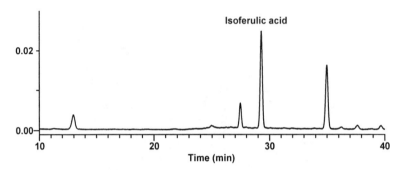

Figure 4. HPLC chromatogram of isoferulic acid in black cohosh extract.

Conventional HPLC-UV methods for the determination of triterpene glycosides are not appropriate due to the low detection wavelength required and the high background noise resulting from phenolic constituents.

Figure 5 shows a chromatogram of a triterpene glycoside fingerprint detected with a conventional HPLC-UV-detector at 203 nm. The signal to noise ratio and peak purity are very poor. Also the baseline is too noisy.

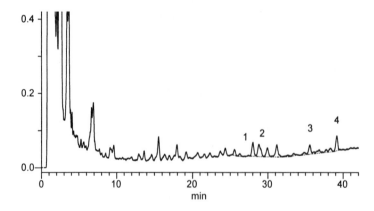

Figure 5. HPLC chromatogram of black cohosh extract, UV-trace at 203 nm.

Actein (1), deoxyactein (2) and both cimicifugosides (3 and 4) were determined as the characteristic constituents and calculated as 27-deoxyactein.

The introduction of the Evaporative Light Scattering Detector (ELSD) significantly improves the chromatographic picture (Figure 6).

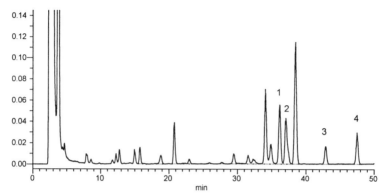

Figure 6. HPLC chromatogram of black cohosh extract, ELSD-trace.

Beside a baseline separation of the triterpene glycosides the sensitivity, selectivity and the signal noise ratio increase significantly. Advantages of Evaporative Light Scattering detection are:

- All compounds can be detected. The presence of no chromophoric group, no electroactive group, etc. is required
- The detector response is directly related to the mass of the eluted compound. Accurate quantitative analytical data can be obtained for unidentified compounds.
- Gradient elution can be used to separate the sample solution. Since the mobile phase is removed from the eluent before detection, a gradient can be used to optimize the separation within a comparatively short time.
- Ideal method for molecules with higher molecular masses like triterpene glycosides.

The disadvantage of Evaporative Light Scattering detection is that the detector signal is not linear. Therefore multilevel calibration within a wider range is essential.

Experimental

Calibration

For calibration a stock solution of 0,2628 mg/ml deoxyactein in methanol was diluted six times to the corresponding concentrations of 52.56, 78.84, 105.12, 131.40, 157.68 and 183.96 μg/mL. The analysis of each of the six calibration samples was done in triplicate.

By drawing the natural logarithm of peak area versus the natural logarithm of the corresponding concentration a linear calibration curve showing a correlation coefficient of 0,9976 was obtained. The linear calibration function was y = 1,8129x + 5,4837.

Sample Preparation

Roots of black cohosh and pharmaceutical preparations were ground to a fine powder in a mortar and weighed (ca. 2 g) into 50 mL flasks. After the addition of 40 mL methanol the samples were placed in an ultrasonic bath for 1 hour and were filtered through Teflon membranes after the volume was made up to 50 mL. Twenty μL of sample were injected into the HPLC system.

Black cohosh extract preparations were directly weighed into the 50 mL flasks and analyzed as described above.

Instrumentation

The analysis was performed using a Shimadzu HPLC system connected to a SEDEX 75 evaporative light scattering detector. For chromatography a Phenomenex Prodigy RP 18 column (250/4.5 mm; 5 μm particle size) was used. The solvent system consists of pure water and pure acetonitrile both obtained in the highest purity available.

Standardization of Black Cohosh Extracts

Due to a comprehensive quality assurance system with control of all steps of the manufacturing process, it is possible to assure herbal drugs and herbal drug preparations with optimum and consistent quality according to the current standard of scientific knowledge. In this respect the efforts have to refer to the reproducibility of the total pattern of constituents and not to individual substances, because in the case of herbal medicinal products the herbal drug preparation in its entirety is regarded as the active substance. By the means of the newly designed ELSD method it is now possible to monitor triterpene glycosides both qualitatively and quantitatively through the entire production process and hence monitor the quality of the herbal drug preparation (Figure 7).

As a result the triterpene fingerprint of the herbal drug can be successfully reproduced in the finished herbal drug preparation. And this is a basic precondition to reproduce quality and hence efficacy of the herbal medicinal product.

Especially on the US-market black cohosh extracts are adjusted ("standardized") to a certain amount of triterpene glycosides, mostly 2.5% triterpene glycosides calculated as 27-deoxyactein. Although pharmaceutically relevant, triterpene glycosides are considered for quality control purposes only and NOT to adjust black cohosh extracts on certain levels. This is because they do not contribute to efficacy. These amounts could fake better efficacy. Further arbitrary increases of triterpene glycoside levels automatically dilute other constituents in the herbal drug preparation, which might possibly contribute to efficacy. A proposal for a standardized black cohosh extract is shown in Table II.

Table II. Specification for Checking the Quality and Batch-Conformity of Black Cohosh Extracts – Main Features Only

Production Data	
Species	*Cimicifuga racemosa* (L.) NUTTAL
Botanical part used	Rhizome (*Cimicifugae rhizoma*)
Extraction solvent	Ethanol 50 % (m/m)
Ratio of herbal drug to drug preparation (= Native extract)	4 – 9 = 1
Composition of the extract preparation	15 % native extract, 85 % Carrier
Quality Data	
Identity	
Thin layer chromatogram	Fingerprint chromatogram on Cimicifuga saponines
Assay (Marker compound)	
HPLC-UV	Isoferulic acid (batch specific)
HPLC-ELSD	Triterpene glycosides (batch specific)

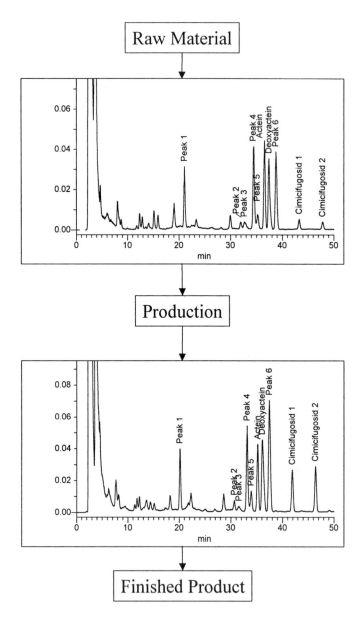

Figure 7. HPLC fingerprint of herbal drug and corresponding herbal drug preparation of black cohosh, ELSD-trace.

A Short Examination of Available Black Cohosh Herbal Medicinal Products

Figure 8 shows some examples of herbal medicinal products currently in the stores.

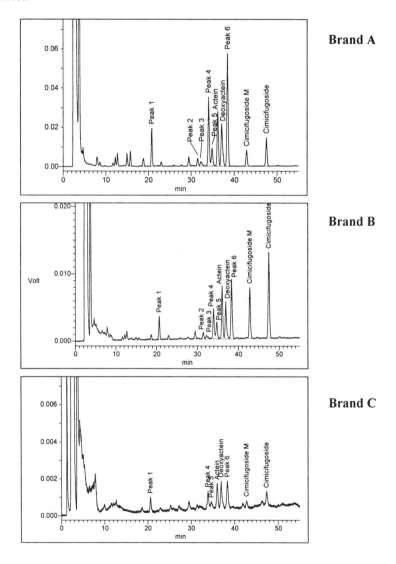

Figure 8.: Comparison of different commercial black cohosh remedies.

Brand A shows the fingerprint of a black cohosh herbal medicinal product being tested in clinical trials. Brand B shows the same profile as A. However, the distribution of the individual triterpene glycoside peaks is different. Brand C claimed to have the same content as Brand B, but it is obvious that the sample contains only traces of triterpene glycosides compared to with the aforementioned product.

In order to show the triterpene glycoside composition of various herbal medicinal products 8 Brands were analyzed for their triterpene glycoside content. See Figure 9.

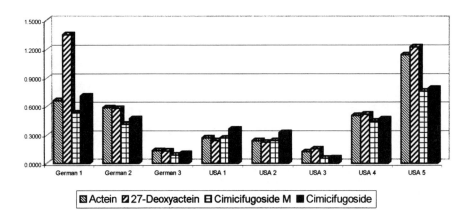

Figure 9. Triterpen glycoside composition in different black cohosh samples.

It becomes obvious that the content and the distribution of the triterpene glycosides varies significantly in different brands. For instance, the German brand 1 shows a very unusual ratio of deoxyactein to the other triterpene glycosides. On the other hand, the American brands 4 and 5 show increased cimicifugoside peaks.

Although the majority of the US black cohosh remedies label a content of 2.5% triterpene glycosides it becomes obvious that different methods are most likely the reason for these variations.

Table III lists a variety of ways to determine triterpene glycoside and their obtained results. Quantification can be achieved spectrophotometrically as well as via HPLC. The values of the widespread and very unspecific photometric method show high results. The HPLC determination with ELSD detection results in lower values, however they are more selective, sensitive and precise. The average content of isoferulic acid in a powdered extract is around 0.2 % (HPLC).

Table III. Ways to Determine Relevant Black Cohosh Constituents

Constituents	Method	Detection	Content
Total triterpene glycosides (after derivatization)	Photometric	UV 460 nm	8.0%
Actein, deoxyactein, cimicifugoside M and cimicifugoside	HPLC	UV 203 nm	2.0%
Total triteroene glycosides including actein, deoxyactein, cimicifugoside M and cimicifugoside	HPLC	ELSD	3.3%
Isoferulic acid	HPLC	UV 320 nm	0.2%

Summary

- The photometric determination of triterpene glycosides is too unspecific.
- The selectivity and reproducibility of HPLC determination of triterpene glycosides with UV-detection is poor.
- The HPLC determination of triterpene glycosides via ELSD is selective and precise
- Isoferulic acid and triterpene glycosides are both possible markers in quality control.

Acknowledgements

The authors gratefully thank the lab team of Finzelberg for running the analytical methods.

References

1. Lieberman, S. *J. Womens Health*, **1998**, *7*, 525-529.
2. Commission E of the Federal Public Health Department: Monograph: Cimicifuga racemosa (Cimicifugawurzelstock), *Bundesanzeiger*. No. 43, dated March 2nd 1989.

3. Liske, E. *Advances in Natural Therapy*, **1998**, *15 (1)*, 45-53.
4. He, K.; Zheng, B.; Kim, C. H.; Rogers, L.; Zheng, Q. *Planta Med.* **2000**, *66*, 635-640.
5. Shao, Y.; Harris, A.; Wang, M.; Zhang, H.; Cordell, G. A.; Bowman, M.; Lemmo, E. *J. Nat.Prod.* **2000**, *63 (7)*, 905-910.
6. Struck, D.; Tegtmeier, M.; Harnischfeger, G. *Planta Medica* **1997**, *63*, 289.
7. McCoy, J.; Kelly, W. ACS National Neeting, Orlando, Florida, Aug 25th – 29th 1996.
8. Jarry, H.; Harnischfeger, G.; Dueker, E. *Planta Med.* **1985**, *4*, 316-319.
9. Bruchhausen, F. von (Ed.) Hagers Handbuch der Pharmazeutischen Praxis Berlin, New York: Springer. 5. Ed., Vol 2. Drogen, A-K; Blaschek, W.; Haensel, R.; Keller, K.; Reichling, J.; Rimpler, H.; Schenider, G., (Ed) Beuscher, N., Cimicifuga. pp 369-381.
10. Gaedcke, F.; Steinhoff, B.; *Phytopharmaka*; Wiss. Verl.-Ges., 2000, pp 23-26.

Chapter 9

Systematic Investigation on Quality Management of Saw Palmetto Products

Tang-Sheng Peng[1,3], William F. Popin[1], and Marlin Huffman[2]

[1]Twinlab Utah Division, 600 East Quality Drive, American Fork, UT 84003
[2]Plantation Medicinals, Inc., 1401 Country Road 830, Felda FL 33930
[3]Current address: Pure World Botanicals, 375 Huyler Street, South Hackensack, NJ 07606

A systematic investigation of botanical and fatty acid profiles of saw palmetto (*Serenoa repens* (Bartr.) Small) products was conducted. Different products were analyzed for the morphological identification and quantification of major fatty acid constituents. The investigation included seed, fruit, fruit powder, powder mixtures and oil extracts, as well as mixtures of saw palmetto and other plant extracts, including pumpkin (*Cucurbita pepo)* seed oil. Each product has a characteristic fatty acid profile that can be used for the identification and standardization of different saw palmetto products. Examples of questioned products are presented.

Saw palmetto (*Serenoa repens* (Bartr.) Small) is one of the most popular herbal products in the U.S. dietary supplement market. It has been continuously listed on the top-ten lists of several surveys with the 1997 sales of more than $18 millions. It is also one of the most frequently prescribed phytomedicines in Germany with the 1996 retail sales of more than $24 millions (*1-5*). It has been estimated that 2.8 million pounds of saw palmetto fruit are shipped to Europe each year for preparation of saw palmetto products as phytomedicines or nutritional supplements (*6*).

Saw palmetto has a long history of various applications (*6-7*). The primary therapeutic applications of saw palmetto products are for treatment of benign prostatic hyperplasia (BPH) and other related disorders, such as urinary tract inflammation (*1-3,8-12*). Treatment of BPH in the US exceeds $2 billion in costs, accounting for 1.7 million physician office visits and results in more than 300,000 prostatectomies a year (*9*). Many studies on the clinical applications of saw palmetto products have been published. A recent literature search revealed at least 58 clinical trials (*11*). Although more systematic studies are needed, many of these studies have showed that saw palmetto extracts (SPE) reduces BPH symptoms with fewer side effects than synthetic drugs like finasteride (*13-15*). The review article listed 10 mechanisms, including the inhibition of 5α-reductase, revealed by many studies to support the polypharmaceutical nature of saw palmetto products (*11*). Importantly, saw palmetto products have no effect on prostate specific antigen (PSA) levels and, therefore, will not mask prostatic carcinoma during a BPH medication.

Most, if not all, of the reviewed clinic trials used lipophilic extract of dried saw palmetto fruit. These extracts can be obtained from saw palmetto fruits through different procedures such as press, solvent extraction, and supercritical carbon dioxide (CO_2) extraction. Most of the SPE suppliers we have worked with use supercritical CO_2 extraction that produces SPE directly without further purification. Approximately 10% of SPE can be obtained from ripened dried saw palmetto fruit. These lipophilic extracts contain a broad brand of chemical components including free fatty acids, fatty alcohols, fatty acid esters, neutral fats, and phytosterols (*12,16*). Although more studies are needed to ascertain the correlation of the chemical components and the pharmaceutical properties of SPE, some researchers have revealed that the free fatty acids and phytosterols in saw palmetto products played active roles in BPH treatments (*17-19*).

The major components of lipophilic SPE are fatty acids and their esters. It is reported that the lipid content of SPE consists of 75% free fatty acids and 25% free fats (*20*). Since other constituents are minor and more difficult to be quantified, total fatty acid content has been selected by the dietary supplement industry as one of the specifications for the quality control of saw palmetto products. In the US dietary supplement market, almost all SPE oils are labeled to contain 85-95% total fatty acids. We have analyzed numerous saw palmetto samples during our quality control process in our laboratories and have found that the fatty acids content specification is not only good for quality control of SPE, but also useful for testing the quality of other saw palmetto products such as saw palmetto fruit powder. During the saw palmetto price crisis in 1995, we received several samples from a supplier labeled as "saw palmetto powder". The consequent analysis showed fatty acid profiles very different from that of authentic saw palmetto fruit. This prompted us to set up a program to investigate systematically the major types of saw palmetto products in the dietary supplement market and the results are reported below. A part of the preliminary results has been communicated earlier (*21*).

Botanical Characteristics of Saw Palmetto Plant and Fruit

Saw palmetto plant (*Serenoa repens* (Bartr.) Small, family Palmae) is a dwarf palm commonly growing wild in Florida and along the southern Atlantic coast (*1, 6-7*). Plantation Medicinals, Inc. has set up an experimental program to cultivate saw palmetto in South America. The plants have been growing vigorously but have not reached the fruiting stage. *Serenoa repens* may attain a height of 2 meters and has creeping, horizontal stems from which asexual reproduction may occur (*1,6-7*). The common name for *Serenoa repens*, saw palmetto, is derived from the leaves which are fan shaped, stiff, and reminiscent of saw blades (*1,6-7*).

The fleshy fruit of *Serenoa repens* occurs as single seeded drupes and both the fruit and seed are processed for use in pharmaceutical and nutraceutical preparations. The immature *Serenoa repens* fruit is green to mottled green-black and turns yellow-orange to black when mature. The fruit turns brown to black when dried. The dry, mature fruit from *Serenoa repens* is ovoid to ellipsoid, 18-30 mm in length and 12-14 mm in width. Polygonal depressions occur in mature, dried *Serenoa repens* fruits. The immature *Serenoa repens* fruit is green and turns brown-black when dried. The shape of immature fruit is ovoid to ellipsoid, around 16.0 mm in length, 10 mm in width, and wrinkled. Deep polygonal depressions are absent in dried immature fruit. The seed of *Serenoa repens* is orange-brown.

A photograph of immature, mid-season semi-mature, and fully mature *Serenoa repens* fruits is shown in Figure 1. These samples were collected by Plantation Medicinals Inc. in a colony approximately 3 meters in diameter near Punta Gorda, Florida, during the 1995 harvest season from August 18th to September 6th. The fresh fruit was separated by hand into three groups according to the maturity stages by color: immature (green), semi-mature (yellow), and fully mature (black). The grouped fruit was then dried in an oven at 55°C for 72 hours and then cooled in oven. The average physical data of *Serenoa repens* fruits were measured from 10 fruits randomly selected from a one (1) pound sample and are summarized in Table I. From the data in Table I, the average weight of fully mature fruit can be up to 1.5 times of that of immature fruit. However, the percentages of flesh and seed are almost identical.

Figure 1. Immature, mid-season, and mature fruits of Serenoa repens.
Reprinted with permission from reference 21. Encyclia 1996.

Table I. Average Physical Data of Saw Palmetto Fruits

Sample Description	Length (mm)	Width (mm)	Weight (mg)	Flesh (%)	Seed (%)
Immature Fruit	16.1	9.9	887	57.6	42.4
Semi-mature Fruit	20.0	11.7	1099	53.1	46.9
Full Mature Fruit	19.1	13.6	1307	57.4	42.6

Although the harvest season can vary from year to year depending on the weather conditions, most of the immature saw palmetto fruit on the market is harvested in late July or early August by collectors who want to make some quick money. Fruit collected earlier is too small to be commercially harvested. The saw palmetto fruit turns deep black and soft in early October when it is fully mature. This makes the harvesting and transportation processes more complicated. Therefore, most of the saw palmetto fruit products are harvested in the mid-season from late August to September when >75% of the fruit are yellow-orange or turning from brown to yellow. The hand-harvested fruit is transported from the wild to the processing plant and then dried at about 55°C for approximately 72 hours. The resulting product is whole dried fruit with a moisture content of 10% or less. The dried and cleaned fruits are stored in controlled conditions before further processing into saw palmetto powders or extracts.

Fatty Acid Profile of Saw Palmetto Fruit and Fruit Powders

Based on the preceding results, *Serenoa repens* fruits, and their maturity at harvest, can be readily identified and distinguished from other "palmetto" species (see below) by the shape, size, weight, color, and external morphology of the fruits. Morphological analysis of whole palmetto fruits utilizing voucher specimens or herbarium sheet standards is a useful method for the preliminary identification of saw palmetto fruits, especially when the product is in the form of whole fruit. However, most of the commercial saw palmetto products are milled into powders or processed into extracts. Therefore, identification based on the chemical components of the products is necessary.

Since fatty acids are the major components of the lipophilic extracts of saw palmetto fruit, we decided to set up a quality control protocol for saw palmetto products using fatty acid profiles. The gas chromatography (GC) method for fatty acid analysis was modified from the AOAC method for oils and fats (22). This modified method has been submitted to USP and published as a part of the official saw palmetto monograph (23).

Thus, the dried saw palmetto fruits were milled to powders (≤ 40 US mesh) with a Handi-Magic mill (Magic Mill). The fatty acids and fatty acid-containing esters were saponified with 0.5 N methanolic NaOH solution and converted to methyl esters in the presence of $BF_3 \bullet MeOH$ (12.5%, Aldrich). The resulting fatty acid methyl eaters were extracted with hexane and analyzed by GC with a 30 m HP-INNOWax capillary column using helium as carrier gas and FID detector. The oven temperature was set at 120°C for 3 minutes and then raised to 220°C in 2 minutes. The total analyzing time is about 15 minutes. Nonadecane (Aldrich) was used as an internal standard. A representative chromatogram is shown in Figure 2 and the fatty acid contents of the saw palmetto fruits at different maturity stages are shown in Table II and Figure 3.

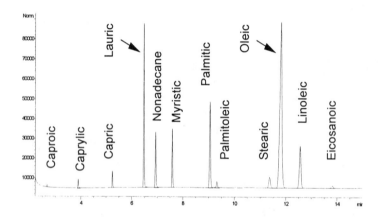

Figure 2. Representative GC chromatogram of the fatty acid methyl esters from the mid-season semi-mature saw palmetto fruit.

All of the saw palmetto fruit samples contained the saturated fatty acids with even carbon numbers from caprylic acid (C-8) to stearic acid (C-18) and unsaturated oleic acid, linoleic acid, and linolenic acid, with lauric and oleic acids as major components followed by myristic and palmitic acids (Table II, Figures 2 & 3). Caproic acid (C-6), eicosanoic acid (C-20), and unsaturated palmitoleic acid (C-16:1) may also be present in a very low concentration. The content of oleic acid increases significantly as the fruit matures, while the content of the other fatty acids remain relatively constant. The total fatty acid content increased from 8.18% in the immature fruit to 10.32% in the fully mature fruit with the semi-mature fruit at 9.41%. Unlike most plant seeds, γ-linolenic acid is absent or present in very low concentration in saw palmetto.

Table II. Fatty Acid Contents of the Saw Palmetto Fruits at Different Maturity Stages

Fatty Acid	Immature Fruit (%)	Semi-mature Fruit (%)	Fully Mature Fruit (%)	Semi-mature Flesh (%)	Semi-mature Seed (%)
Caprylic	0.08	0.12	0.16	0.09	0.20
Capric	0.16	0.19	0.21	0.10	0.41
Lauric	2.59	2.80	2.46	4.02	2.50
Myristic	1.11	1.22	1.15	2.00	0.57
Palmitic	0.90	0.94	1.11	1.23	0.71
Stearic	0.20	0.22	0.21	0.16	0.32
Oleic	2.54	3.31	4.57	4.09	2.54
Linoleic	0.50	0.51	0.37	0.19	1.06
Linolenic	0.10	0.10	0.08	0.15	0.00
Total	8.18	9.41	10.32	12.03	8.31

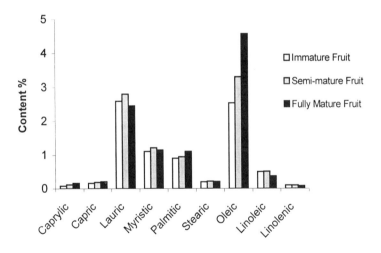

Figure 3. Fatty acid profiles of immature, mid-season, and mature saw palmetto fruits.

The flesh of saw palmetto fruit contributed more to all four major fatty acids (oleic, lauric, myristic, and palmitic acids) but the seed may contribute more to some of the minor components such as linoleic and capric acids (Figure 4). The flesh contains 45% more total fatty acids than the seed. In the dietary supplement industry, some saw palmetto products are sold as "cut" form – the whole fruits are cut up to pieces with a diameter of ca. 5 mm. This process separates seed from the flesh part of the fruit. Since the density of seed is much higher than flesh, a part of the batch may contain more seed fragments and the other part may contain more flesh pieces therefore, it renders the quality of the products less consistent.

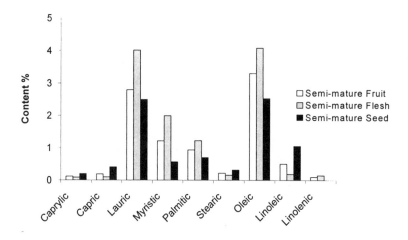

Figure 4. Fatty acid profiles of the fruit, flesh, and seed of a mid-season saw palmetto fruit.

However, most of the saw palmetto fruits are milled into powders with different particle sizes for making capsules or for further processing to make other forms such as extracts. Most of the commercial saw palmetto powders we analyzed had fatty acid profiles similar to that of the authentic saw palmetto fruit. However, in 1995 harvest season when the price for fresh saw palmetto rocketed from $0.13 per pound to $2.34 per pound (6), we received several commercial powder samples labeled as "saw palmetto". GC analysis of these samples revealed low fatty acid contents or fatty acid profiles different from that of the authentic saw palmetto fruit. Representative results are shown in Table III and Figure 5 using the semi-mature saw palmetto as an authentic standard. Sample 1 had a fatty acid distribution similar to the authentic saw palmetto fruit but the contents were low for major fatty acids. Samples 2 to 4 had fatty acid profiles very different from the authentic saw palmetto. Reasons for this include the use of immature fruit or diluted material (for sample 1), different "palmetto" species (see below), or spent material with addition of non-saw palmetto oils (see below).

Table III. Fatty Acid Contents of the Authentic Saw Palmetto Fruit and Questionable Commercial Samples

Fatty Acid	Authentic (%)	Commercial 1 (%)	Commercial 2 (%)	Commercial 3 (%)	Commercial 4 (%)
Caprylic	0.12	0.00	0.00	2.30	1.90
Capric	0.19	0.17	0.02	1.20	1.00
Lauric	2.80	1.90	0.22	0.20	0.10
Myristic	1.22	0.82	0.10	0.00	0.00
Palmitic	0.94	0.71	1.30	0.08	0.30
Stearic	0.22	0.15	0.57	0.03	0.10
Oleic	3.31	2.70	3.50	0.54	1.00
Linoleic	0.51	0.49	6.10	0.20	1.20
Linolenic	0.10	0.05	0.63	0.00	0.10
Total	9.41	6.99	12.44	4.55	5.70

Other "Palmetto" Species

Serenoa repens, the sole representative of this genus (*6*), is also known by its synonyms including *Sabal serrulata* (Michaux) Nuttall. Other *Sabal* and Palmae species growing from south USA to South America are also called "palmetto" by local residents. Although the other palmettos are distinct from saw palmetto to trained saw palmetto harvesters, we have seen powdered "*Sabal*" fruits on the market during some harvest seasons. Therefore, it is necessary for people engaged in saw palmetto product manufacturing and consumers to know the possible adulterants.

The two most commonly substituted Palmae species are *Sabal minor* (Jacq.) Persoon (bush palmetto), and queen palm *Syagrus romanzoffiana* (Cham.) Glassm. *Sabal minor* occurs as a bush or dwarf palm 2-3 meters in height with fan shaped leaves to 1.0-1.5 meters and its fruit is a single seeded drupe that turns brown to black at maturity (*24*). The dry fruit from *Sabal minor* is spherical (globe shaped) 10-12 mm in diameter and polygonal depressions rarely occur. The seed of *Sabal minor* is dark red. Compared to the fruit of *Serenoa repens*, the fruit of *Sabal minor* is smaller in size and lighter in weight (see Table IV). *Syagrus romanzoffiana* may attain a height of 10-20 meters with pinnate leaves to 5 meters and the fruit of *Syagrus romanzoffiana* is light brown to brown. The dry fruit from *Syagrus romanzoffiana* is ovoid, 20-25 mm in length and 13-15 mm in width. The fruit is smooth to wrinkled and absent of polygonal depressions. Compared to the fruit of *Serenoa repens*, the fruit of *Syagrus romanzoffiana* is significantly larger and heavier (Table IV). Photographs of representative *Sabal minor* and *Syagrus romanzoffiana* fruits harvested in September 1995 near Punta Gorda of Florida are shown in Figure 6.

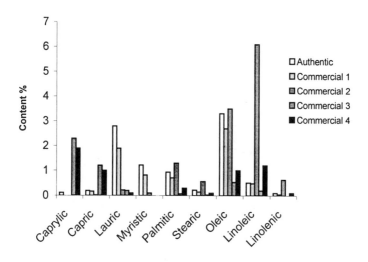

Figure 5. Fatty acid profiles of the authentic and questionable commercial saw palmetto products.

Table IV. Average Physical Data of the fruits of Serenoa repens and Related Species

Sample Description	Length (mm)	Width (mm)	Weight (mg)
Serenoa repens	20.0	11.7	1099
Sabal minor	11.9	10.9	330
Syagrus romanzoffiana	23.9	14.4	2359

Without spiking of foreign oils, powders from *Sabal minor* and *Syagrus romanzoffiana* fruits are less oily and less aromatic than the crude powders from *Serenoa repens* fruit. GC analytical results shown in Table V and Figure 7 revealed that the fatty acid contents of the fruits of *Sabal minor* and *Syagrus*

Sabal minor fruit

Queen palm fruit

Figure 6. Photographs of the fruits of Sabal minor and Syagrus romanzoffiana. Reprinted with permission from reference 21. Encyclia 1996.

romanzoffiana are much lower than those of *Serenoa repens* products. Unlike saw palmetto, palmitic acid content in *Sabal minor* fruit is higher than the contents of lauric acid and myristic acid. Although oleic acid is also the major fatty acid in *Sabal minor fruit*, its content is much lower than that in *Serenoa repens*. Similar to *Serenoa repens*, *Syagrus romanzoffiana* fruit has lauric and oleic acids as major fatty acid components, their contents are less than half of those in *Serenoa repens*. Unlike *Serenoa repens* and *Sabal minor*, the palmitic acid content is higher than myristic acid in *Syagrus romanzoffiana*. These differences may help concerned parties properly identify *Serenoa repens* products.

Table V. Comparison of Fatty Acid Contents of the Fruit of Serenoa repens and Related Species

Fatty Acid	Serenoa repens Fruit (%)	Sabal minor Fruit (%)	Syagrus romanzoffiana Fruit (%)
Caprylic	0.12	0.00	0.19
Capric	0.19	0.00	0.22
Lauric	2.80	0.23	1.24
Myristic	1.22	0.12	0.34
Palmitic	0.94	0.34	0.66
Stearic	0.22	0.10	0.17
Oleic	3.31	1.27	1.31
Linoleic	0.51	0.52	0.31
Linolenic	0.10	0.07	0.00
Total	9.41	2.65	4.44

Fatty Acid Profile of Saw Palmetto Extracts and Their Finished Products

Besides the tea-cut and powdered products processed directly from saw palmetto fruit, most of the other saw palmetto products are made from saw palmetto fruit extracts and standardized for certain percentages of fatty acids. The representative data from the GC analysis of the extract products are summarized in Table VI. A supercritical CO_2 extract of *Serenoa repens* fruit, received from Botanicals International, is a yellow to yellowish oil that is a 90% fatty acid content. The color of an ethanol extract, also an oil, received from Botanicals International, is greenish in color, and easily differentiated from the

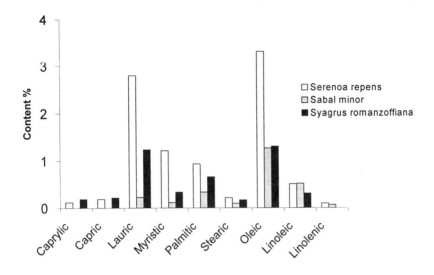

Figure 7. Fatty acid profiles of the authentic saw palmetto and other "palmetto" fruits.

Table VI. Fatty Acid Contents of Saw Palmetto Fruit and Extracts

Fatty Acid	Fruit (%)	Powder Extract (%)	CO_2 Oil Extract (%)	EtOH Oil Extract (%)
Caprylic	0.12	0.40	1.50	1.55
Capric	0.19	0.50	2.10	2.48
Lauric	2.80	9.30	27.00	24.86
Myristic	1.22	4.20	10.80	11.50
Palmitic	0.94	3.20	8.10	9.29
Stearic	0.22	0.60	1.70	1.61
Oleic	3.31	10.70	31.00	33.30
Linoleic	0.51	1.30	5.60	2.90
Linolenic	0.10	0.00	1.60	0.81
Total	9.41	30.20	89.40	88.30

130

CO_2 extract, but the total fatty acid content is very similar to that of the latter. A dried ethanol extract of saw palmetto fruit based on a starch carrier, also from Botanicals International, is a yellowish powder that contains about 30% of fatty acids. Understandably, the total fatty acid contents of the extract products are much higher than that of the crude powder or whole saw palmetto fruit. However, the fatty acid distribution patterns are very similar and proportional for all product forms (Figure 8).

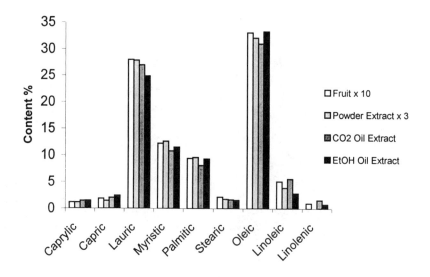

Figure 8. Fatty acid profiles of saw palmetto fruit and fruit extracts

On the finished product market, SPE can be packaged directly to the dosage form or formulated with other ingredients. For instance, saw palmetto oil extract can be sold as soft gel capsules with a daily dosage of 320 mg per capsule, or to be formulated with pumpkin seed oil extract (PSE), another ingredient with BPH treatment effects (*25*). It is easy to monitor the quality of the product with sole saw palmetto extract using the same protocols detailed earlier. However, other ingredients, especially those containing fatty acids such as pumpkin seed oil, may alter the fatty acid profiles of the polyherbal saw palmetto products. It is, therefore, necessary to establish specifications for products with multiple ingredients.

A popular commercial saw palmetto formulation is the 1:1 mixture of saw palmetto oil extract and pumpkin seed oil extract, which has shown effectiveness in reducing BPH symptoms in a clinical study (*26*). Other combinations such as

2:3 formulations of saw palmetto oil extract and pumpkin seed oil extract are also present on the market. Thus, a pumpkin seed oil extract, received from Soft Gel Technologies, is analyzed for fatty acid profile and the result is compared with that of the saw palmetto CO_2 oil extract in Table VII. Similar to SPE, pumpkin seed oil extract also contains about 90% of total fatty acids. However, the fatty acids distribution pattern is very different from that of SPE. PSE contains no or very little lauric and myristic acids, two of the major fatty acids in SPE. Although both SPE and PSE contain palmitic and oleic acids, PSE contains a much higher concentration (almost 50%) of the unsaturated linoleic acid. Thus, a 1:1 SPE/PSE formula should contain an average concentration of each fatty acid contributed by both SPE and PSE. GC results of a 1:1 SPE:PSE mixture and a 2:3 SPE:PSE mixture are shown in Table VII and Figure 9. Soft gelatin capsules with the same SPE/PSE combinations, received from Soft Gel Technologies, yielded similar results.

Table VII. Fatty Acid Contents of Saw Palmetto Finished Products

Fatty Acid	Saw Palmetto Oil Extract (%)	Pumpkin Seed Oil Extract (%)	1:1 SPE:PSE Formula (%)	2:3 SPE:PSE Formula (%)
Caprylic	1.50	0.00	0.74	0.69
Capric	2.10	0.00	1.49	1.32
Lauric	27.00	0.00	14.77	12.72
Myristic	10.80	0.13	6.14	5.23
Palmitic	8.10	12.98	10.18	10.47
Stearic	1.70	5.48	3.47	3.62
Oleic	31.00	26.45	25.48	23.07
Linoleic	5.60	47.54	25.46	31.58
Linolenic	1.60	0.36	0.55	0.51
Total	89.40	92.94	88.28	89.21

In summary, we have conducted a systematic investigation of saw palmetto and related species to identify important morphological and chemical components. Included in our investigation were samples of saw palmetto fruit with different maturity states, dosage forms, commercial products, and fruit samples from other palm species. The fruit and phytopharmaceutical products of saw palmetto contained a characteristic morphology and fatty acid profiles. These properties are distinguishable from other Palmae species and can be utilized for the quality management of saw palmetto raw materials and finished products.

132

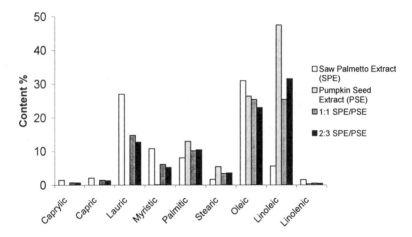

Figure 9. Fatty acid profiles of different saw palmetto dosage forms

Acknowledgments

The authors express appreciation to Douglas F. Johnston, W. Aaron Broyles, and Timothy Johnson who worked on the laboratory analysis and preliminary data collection of saw palmetto products at Twinlab Utah Division. The authors want to thank Richard Holdaway for preparing some of the photographs, and to Daniel Brennan for review of this manuscript. The authors recognize Plantation Medicinals and Twinlab Utah Division for financial and resource support.

References

1. Foster, S.; Duke, J. A. *A field Guide to Medicinal Plants*; Houghton Mifflin Company: Boston, MA, 1990.
2. Duke, J. A. *CRC Handbook of Medicinal Herbs*; CRC Press: Boca Raton, FL, 1985.
3. Tyler, V. E. *Herbs of Choice: The Therapeutic Use of Phytochemicals*; Pharmaceutical Products Press: New York, NY, 1994.

4.   Richman, A.; Witkowski, J. P. *Whole Foods* **1997**, *Oct.* 20-28.

5.   *The Complete German Commission E Monographs*; Blumenthal, M.; Busse, W. R.; Goldberg, A.; Gruenwald, J.; Hall, T.; Riggins, C. W.; Rister, R. S., Eds; American Botanical Council: Austin, TX; Integrative Medicine Communications: Boston, MA, 1998; pp 11-25.

6.   Bennett, B. C.; Hicklin, J. R. *Economic Botany* **1998**, *52*, 381-393.

7.   *United States Dispensatory, 22nd Edition;* Wood, H. C.; LaWall, C. H., Eds.; J. B. Lippincott Company: Philadelphia, PA, 1940.

8.   Lowe, F. C.; Ku, J. C. *Urology* **1996**, *48*, 12-20.

9.   Wilt, T. J.; Ishani, A; Stark, G.; MacDonald, R.; Lau, J.; Mulrow, C. *JAMA* **1998**, *280*, 1604-1609.

10. Plosker, G. L.; Brogden, R. N. *Drugs & Aging* **1996**, *9*, 379-395.

11. McPartland, J. M.; Pruitt, P. L. *JAOA* **2000**, *100(2)*, 89-96.

12. Bombardelli, E.; Morazzoni, P. *Fitoterapia* **1997**, *LXVIII(2)*, 99-113.

13. Carraro, J. C.; Raynaud, J. P.; Koch, G.; Chisholm, G. D.; DiSilverio, F.; Teillac, P. *Prostate* **1996**, *29*, 231-240.

14. Sokeland, J.; Albrecht, J. *Urology[A]* **1997**, *36*, 327-333.

15. Marks, L. S., et al. *J. Urol.* **2000**, *163*, 1451-1456.

16. Jommi, G; Verotta, L. *Gazzetta Chimica Italiana* **1988**, *118*, 823-826.

17. Niederprum, H. J.; Schweikert, H. U.; Zanker, K. S. *Phytomedicine* **1994**, *1*, 127-133.

18. Berges, R. R.; Windeler, J.; Trampisch, H. J.; Senge, T.; the β-Sitosterol Study Group *Lancet* **1995**, *345*, 1529-1532.

19. Klippel, K. F.; Hiltl, D. M.; Schipp, B, *Br. J. Urol.* **1997**, *80*, 427-432.

20. Harnischfeger, G.; Stolze, H. *Z. Phytotherapie*, **1989**, *10*, 71.

21. Peng, T.-S.; Johnston, D. F.; Popin, W. F.; Fralick, G. K, Jr. *Encyclia*, **1996**, *73*, 205-214.

22. *Official Methods of Analysis of AOAC International, 16th ed.*; Cunniff, P., Ed.;   AOAC International: Gaithersburg, MD, 1995; Methods 969.33 and 963.22.

23. *USP 24/NF 19*; United States Pharmacopeial Convention: Rockville, MD, 2000; pp 2510-2512.

24. Griffiths, M *Index of Garden Plants*; Timber Press: Portland, OR, 1994.

25. Bombardelli, E.; Morazzoni, P. *Fitoterapia* **1997**, *LXVIII(4)*, 291-302.

26. Carbin, B. E.; Larson, B.; Lindhal, O. *British J. Urol.* **1990**, *66*, 639.

Chapter 10

Chemical Components in Noni Fruits and Leaves (*Morinda citrifolia* L.)

Shengmin Sang[1], Mingfu Wang[1], Kan He[2], Guangming Liu[3],
Zigang Dong[3], Vladimir Badmaev[4], Qun Yi Zheng[2], Geetha Ghai[1],
Robert T. Rosen[1], and Chi-Tang Ho[1]

[1]Department of Food Science and Center for Advanced Food Technology,
Rutgers University, 65 Dudley Road, New Brunswick, NJ 08901–8520
[2]Research and Development, Pure World Botanicals, Inc., Huyler Street,
South Hackensack, NJ 07606
[3]Hormel Institute, University of Minnesota, Austin, MN 55912
[4]Sabinsa Corp., 121 Ethel Road, Piscataway, NJ 08854

Morinda citrifolia (Rubiaceae), commonly known as noni, is a
plant typically found in the Hawaiian and Tahitian islands. The
bark, stem, root, leaf, and fruit have been used traditionally as
a folk remedy for many diseases including diabetes,
hypertension, and cancer. In this research, we reported the
constituents of the fruits and leaves of this plant from both
Hawaii and India. Several new glycosides and iridoids were
identified. Four compounds isolated from noni fruits were
found to suppress UVB-induced AP-1 activity in cell cultures.

Morinda citrifolia (Rubiaceae), commonly known as noni, is a plant
typically found in the Hawaiian and Tahitian islands. It is believed to be one of
the most important plants brought to Hawaii by the first Polynesians (*1*). The
plant is a small evergreen tree growing in the open coastal regions and in forest
areas up to about 1300 feet above the sea level. This plant is identifiable by its

straight trunk, large green leaves and its distinctive, ovid, "grenade-like" yellow fruit. The fruit can grow to a size of 12 cm and results from coalescence of the inferior ovaries of many closely packed flowers. It has a foul taste and a soapy smell when ripe. The bark, stem, root, leaf, and fruit have been used traditionally as a folk remedy for many diseases including diabetes, hypertension, and cancer (*2,3*). The fruits of this plant were also used as foods in time of famine, whereas the roots were used to produce a yellow or red dye for cloth.

Previous Chemical Studies

In earlier studies, ricinoleic acid was found in the seeds (*4*), whereas anthraquinones including morenone 1, morenone 2, damnacanthal and 7-hydroxy-8-methoxy-2-methylanthraquinone (Figure 1) have been identified in the root of noni (*5,6*). From the heartwood, two known anthraquinones (morindone and physcion) and one new anthraquinone glycoside (physcion-8-*O*-α-L-arabinopyranosyl(1→3)-[β-D-galacotopyranosyl(1→6)]-β-D-galacto-pyranoside) have been isolated from the heartwood of *Morinda citrifolia* (Figure 1) (*7*). Studies on the chemical components of the flowers of noni have resulted in the identification of one anthraquinone glycoside and two flavone glycosides from the flowers of noni (*8,9*) (Figure 2), whereas β-sitosterol and ursolic acid have been isolated from the leaves (*10*). Although the fruits of noni have been used as a food, very few reports on the chemical components of the fruits are available (*1*). Several nonvolatile compounds including acetyl derivatives of asperuloside, glucose, caproic acid, and caprylic acid have been identified in fruits (*1*) (Figure 3). Recently, about 51 volatile components were identified by GC-MS from the ripe fruit (*11*).

Previous Pharmacological Studies

The juice of noni fruits has been shown to prolong the life span of mice implanted with Lewis lung carcinoma (*2*). At a dose of 15 mg noni juice per mouse (original fruit juice=15 mg solid per 0.2 mL volume), noni could significantly increased the life span of C57BL/6 mice implanted with Lewis lung carcinoma by 119%, with 9 out of 22 mice surviving for more than 50 days. And at lower dose, noni fruit juices still showed activity. It was proposed that the fruits of noni might suppress the growth of tumors by stimulating the immune system (*2*). Later, the same authors provided evidence supporting their hypothesis and partly identified a polysaccharide as an active component (*3,12*). They believed that this polysaccharide is composed of the sugars, glucuronic

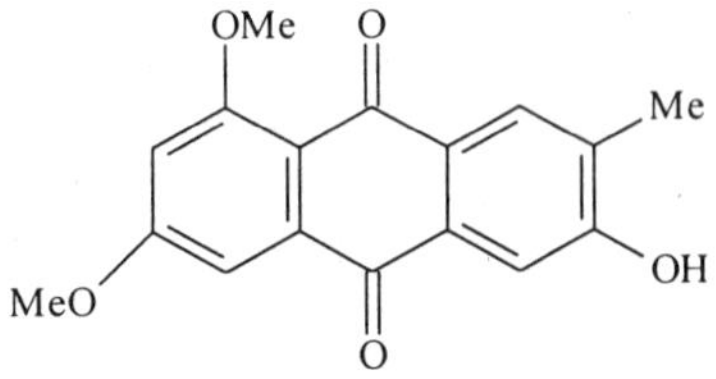

Figure 1. Anthraquinones reported from the seeds and heartwood of Morinda citrifolia.

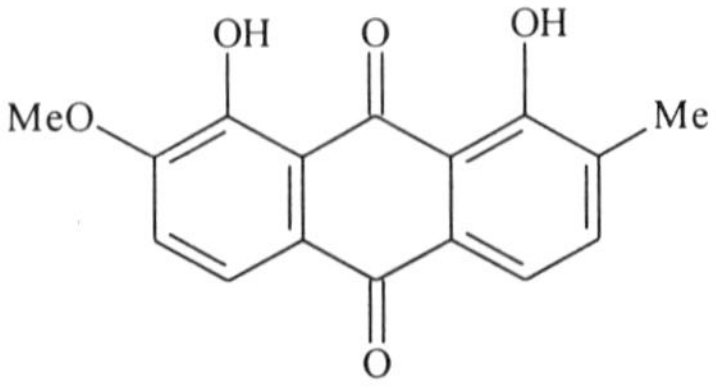

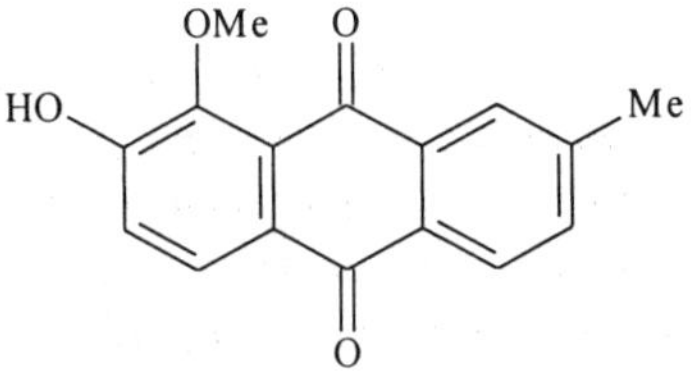

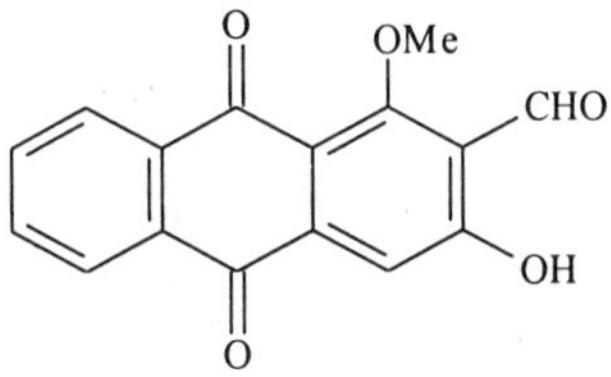

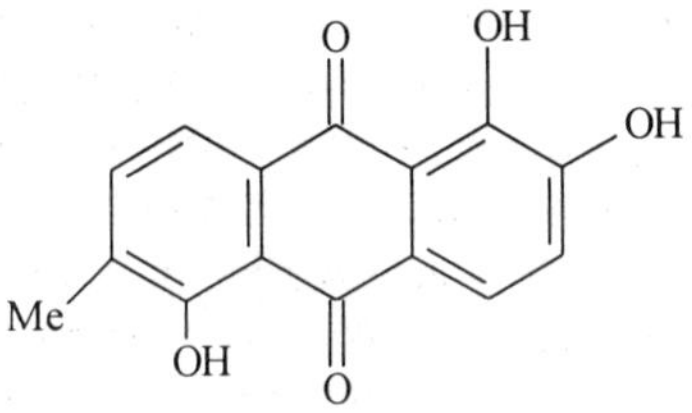

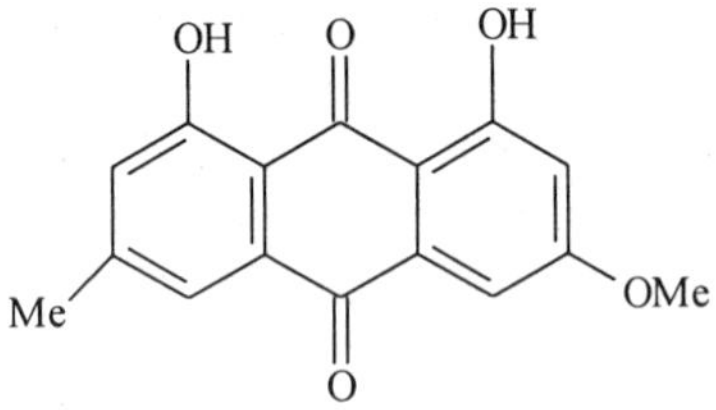

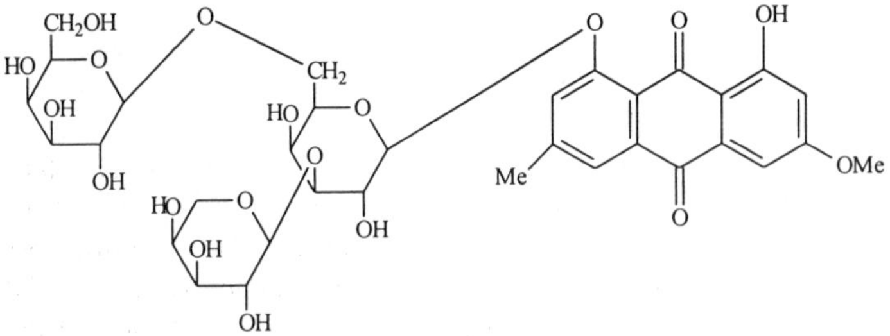

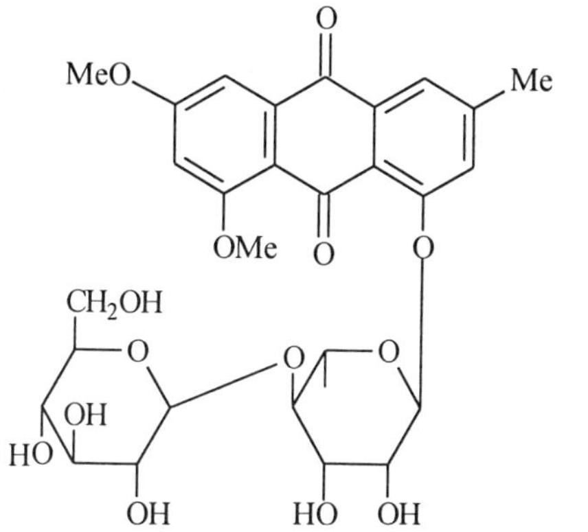

2-methyl-4-hydroxy-5,7-dimethoxyanthraquinone 4-O-β-D-
glucopyranosyl(1→4)-α-L-rhamnopyronoside

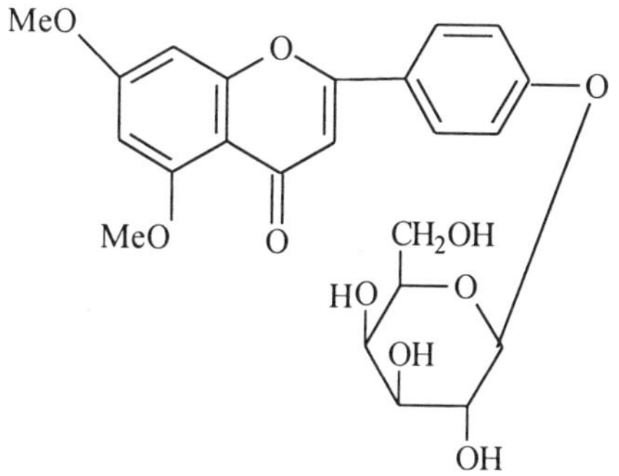

5,7-Dimethyl-apigenin-4'-β-D-galactopyranoside

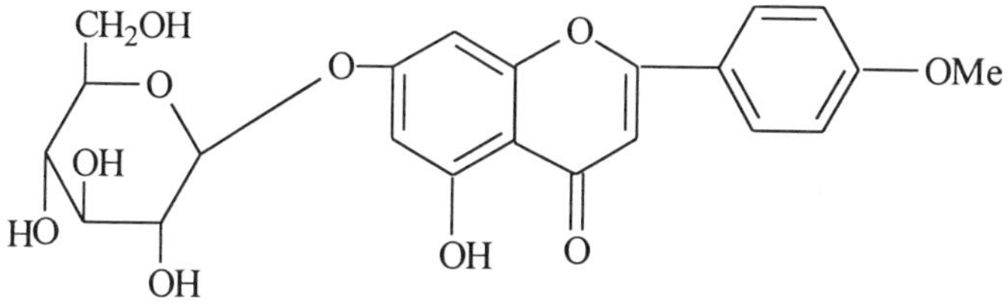

Acacetin-7-O-β-D-glucopyranoside

Figure 2. Anthraquinone glycoside and flavone glycosides identified from the flowers of Morinda citrifolia.

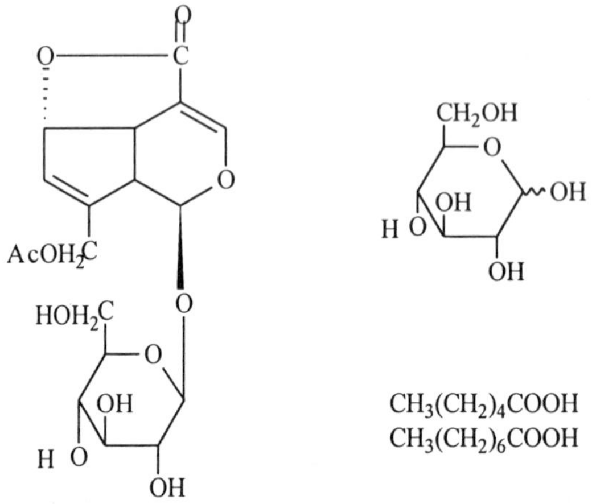

138

Figure 3. Acetyl derivative of asperuloside, glucose and fatty acids reported in the fruits of Morinda citrifolia.

acid, galactose, arabinose and rhamnose with a possible high molecular weight (10000-50000 Daltons). Individual compounds from noni have also been tested for antitumor activity. Damnacanthal, an anthraquinone isolated from the chloroform extract of the roots of noni, has been found to be a new inhibitor of *ras* function and to help to suppress the activated *ras*-expressing tumors. This compounds induced normal morphology and cyoskeletal structure in K-ras[ts]-NRK cells at the permissive temperature, without changing the amount and location of Ras. The effect of this compound was reversible (*13*). In addition, damnacanthal has been reported to inhibit specifically tyrosine kinases such as Lck, Src, Fyn and ErbB-2 with half maximum inhibition concentrations between 17 and 700 nM (*14*). It also showed a unique activity to induce the activation of ERKs and the most potent stimulatory effect on UV-induced apoptosis among the kinase inhibitors examined. This might be due to the specific tyrosine kinase inhibitor activity of damnacanthal. Alternatively, damnacanal might possess an additional activity other than the inhibition of protein kinases (*14*). The extracts of noni roots have also been found to possess a significant, dose-dependent, central analgesic activity in the treated mice (*15*).

Because limited reports on the fruits and leaves of noni were available in the literature, we report here the chemical investigation on the fruits and leaves of this plant. We also compare the constituents of two kinds of noni` fruits: Hawaiian noni fruits and Indian noni fruits.

Materials and Methods

Chemicals

Silica gel (130-270 mesh), Sephadex LH-20 (Sigma Chemical Co., St. Louis, MO) and Lichroprep RP-18 column were used for column chromatography. All solvents used for chromatographic isolation were analytical grade and purchased from Fisher Scientific (Springfield, NJ).

General Procedures

^{1}H NMR and ^{13}C NMR spectra were obtained on a VXR-200 instrument (Varian Inc., Palo Alto, CA), operating at 200 and 50 MHz, or on a VXR-400 (Varian Inc.), operating at 400 and 100 MHz, or on a U-500 instrument (Varian Inc.), operating at 500 and 125 MHz, respectively. Compounds were analyzed in CD$_3$OD and DMSO-d_6 with tetramethylsilane (TMS) as an internal standard. ^{1}H-^{1}H COSY, NOESY, HMQC, and HMBC were performed on a VXR-400 (Varian Inc.), or on a U-500 instrument (Varian Inc.). FAB mass spectra were recorded on a Finnigan MAT-90 instrument. APCI MS was obtained on a Fisons/VG Platform II mass spectrometer. FT-IR spectra were obtained with a Perkin-Elmer 1600 apparatus. Thin-layer chromatography was performed on Sigma-Aldrich TLC plates (250 μm thickness, 2-25 μm particle size), with compounds visualized by spraying with 5% (v/v) H$_2$SO$_4$ in ethanol solution.

Results and Discussion

Analysis of Hawaiian Noni Fruits

As shown in Figure 4, Hawaiian noni fruits were freeze-dried. The dried noni fruit (200 g) pieces were extracted with 95% ethanol (1 L) at room temperature for 1 week. The extract was concentrated to dryness under reduced pressure, and the residue was suspended in water (500 mL) and partitioned successively with hexane (3 × 500 mL), ethyl acetate (3 × 500 mL), and *n*-butanol (3 × 500 mL). The *n*-butanol-soluble fraction (40 g) was subjected to column chromatography on 500 g of silica gel and eluted with ethyl acetate/methanol/water as eluent with increasing methanol and water content

200g dried noni fruits

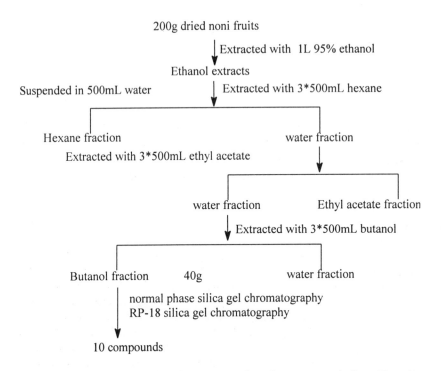

Figure 4. Extraction and isolation procedure for compounds from Hawaiian noni fruits.

(10:1:1, 6:1:1, 6:1:1, 4:1:1, 3:1:1, 2:1:0.5, each 1000 mL), and 500 mL fractions were collected. Fractions 9 and 10 were combined together and rechromatographed on a Lichroprep RP-18 column using methanol/water (1:3, 1000 mL; 2:3, 1000 mL) and methanol to obtain pure compounds **NB-1** (1 g) and **NB-2** (300 mg). Fraction 7-8 were combined together and subjected to a RP-18 column, eluted by methanol:water (1:1) to get the pure compounds 300 mg of **NB-3** and 100mg of **NB-4**. Fraction 6 was first subjected to a normal phase silica gel column, eluted by chlorofrom-methanol-water (3:1:0.1) to get two subfractions, then subfraction 1 was subjected to a reverse phase column eluted with methanol-water (1:1) to get 200 mg of pure **NB-10**. Fraction 4 was subjected to a Sephadex LH-20 column (eluted with methanol) to get two fractions; the first fraction was then subjected to a Lichroprep RP-18 column eluted with methanol/water (1:3, 1000 mL, 2:3, 1000 mL) to give five subfractions, and then subfraction 1 was subjected to a silica gel column using ethyl acetate/methanol/water (10:1:1) as eluent to obtain 100 mg of compound **NB-7**. The second fraction was subjected to a silica gel column using ethyl acetate/methanol/water (7:1:1) to yield 100 mg of compound **NB-11**. Figure 5

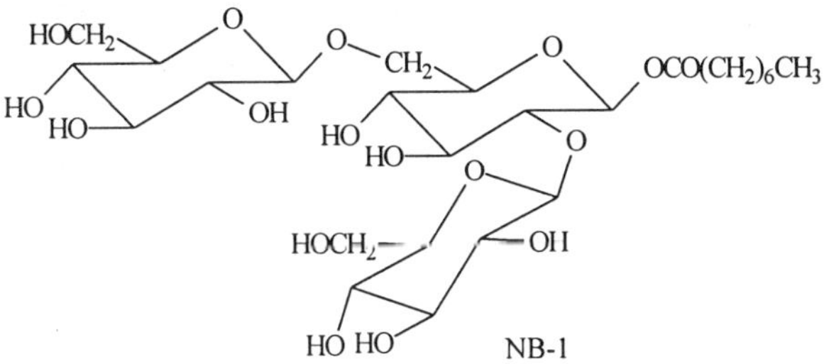

Figure 5. New compounds identified from the Hawaiian noni fruits.

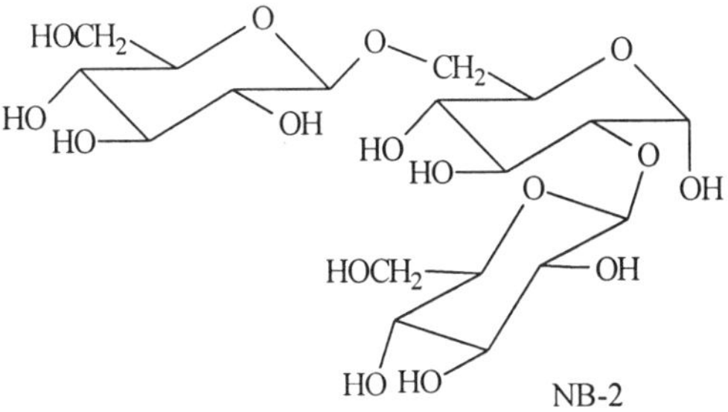

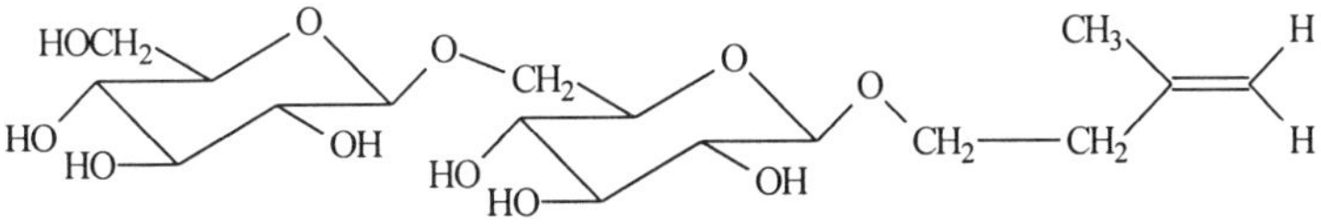

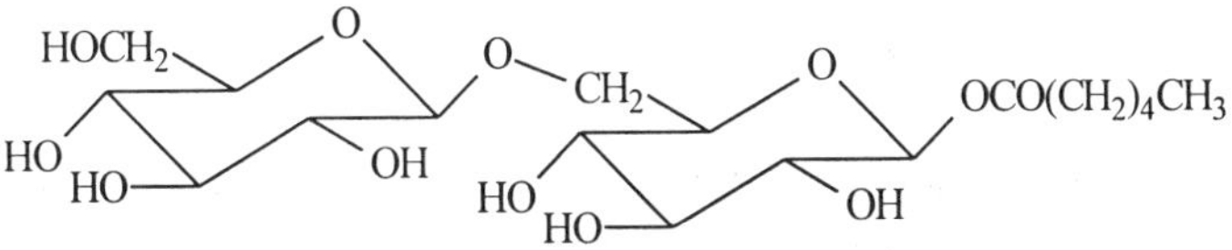

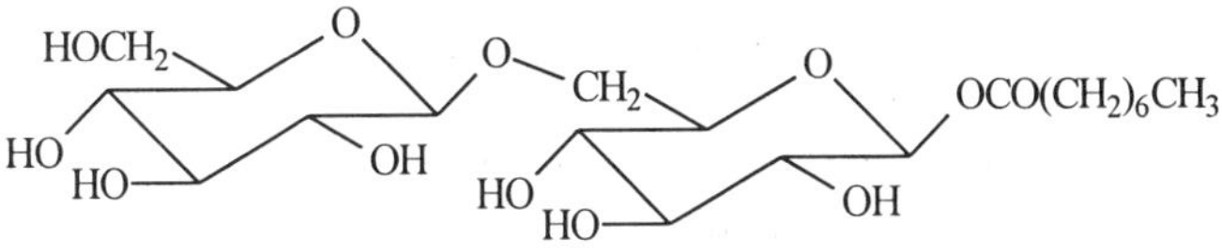

shows the structures of new compounds, **NB-1**, **NB-2**, **NB-3**, **NB-4** and **NB-10** isolated and identified in Hawaiian noni fruits (*16-17*). Figure 6 shows the structures of known compounds, **NB-11** and **NB-7** identifed in Hawaiian noni fruits.

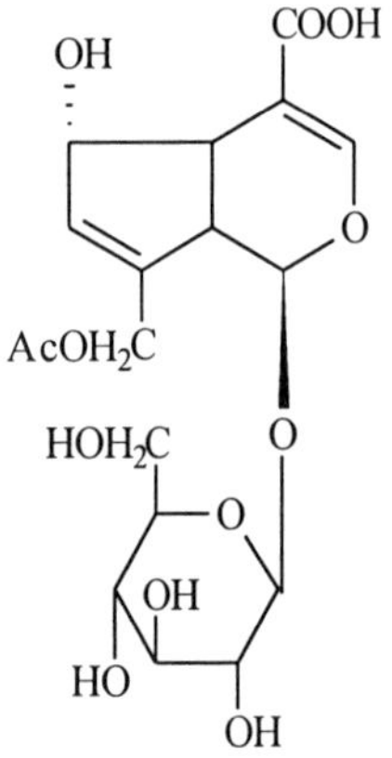

Figure 6. Two known compounds identified from the Hawaiian noni fruits

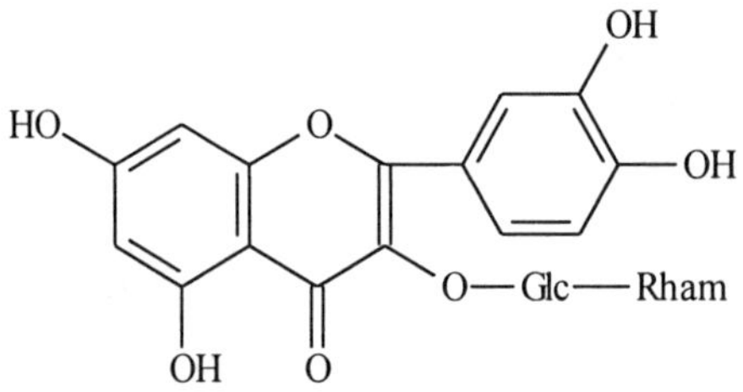

Analysis of Indian Noni Fruits

As shown in Figure 7, the dried Indian noni fruit (5 Kg) powders were extracted with 95% ethanol (4 L) at 50°C for 1 day. The extract was concentrated to dryness under reduced pressure, and the residue was suspended in water (500 mL) and partitioned successively with hexane (3 × 500 mL), ethyl acetate (3 × 500 mL), and *n*-butanol (3 × 500 mL). The butanol fraction was subjected to a Diaion HP-20 column, eluted with water-ethanol (water, 30% ethanol, 70% ethanol, 95% ethanol) solvent system. Four subfractions were obtained. From the ethyl acetate fraction, we obtained two compounds **NFE-1** and **NFE-2**. From the 30% ethanol fraction, 8 compounds have been isolated but only 7 of them were identifiable. Only one out of these 10 compounds identified is a novel compound. Figure 8 shows the structure of this new iridoid compound (**NFB-3**). Figure 9 shows the structures of known compounds identified in Indian noni fruits, they all have been isolated from the noni fruits for the first time.

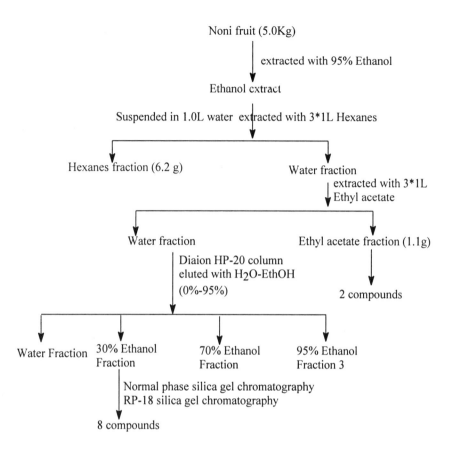

Noni fruit (5.0Kg)

extracted with 95% Ethanol

Ethanol extract

Suspended in 1.0L water extracted with 3*1L Hexanes

Hexanes fraction (6.2 g)   Water fraction

extracted with 3*1L
Ethyl acetate

Water fraction   Ethyl acetate fraction (1.1g)

Diaion HP-20 column
eluted with H_2O-EthOH
(0%-95%)

2 compounds

Water Fraction   30% Ethanol Fraction   70% Ethanol Fraction   95% Ethanol Fraction 3

Normal phase silica gel chromatography
RP-18 silica gel chromatography

8 compounds

Figure 7. Extraction and isolation procedure for compounds from Indian noni fruits.

NFB-3

Figure 8. New compound from the Indian noni fruits.

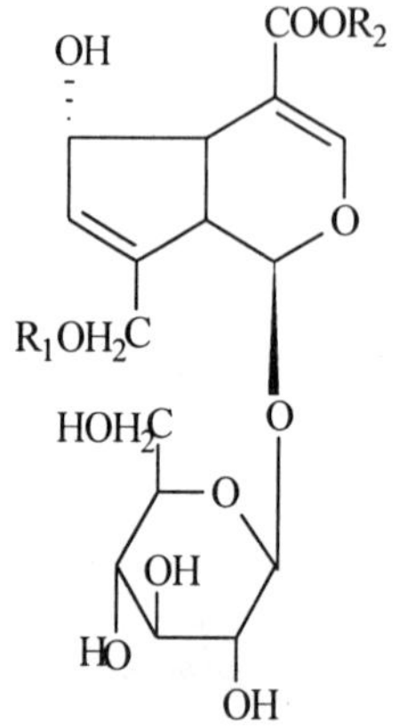

NFB-1: R_1=Ac, R_2=CH_3
NFB-2: R_1=H, R_2=CH_3
NFB-5: R_1=Ac, R_2=H
NFB-7: R_1=R_2=H

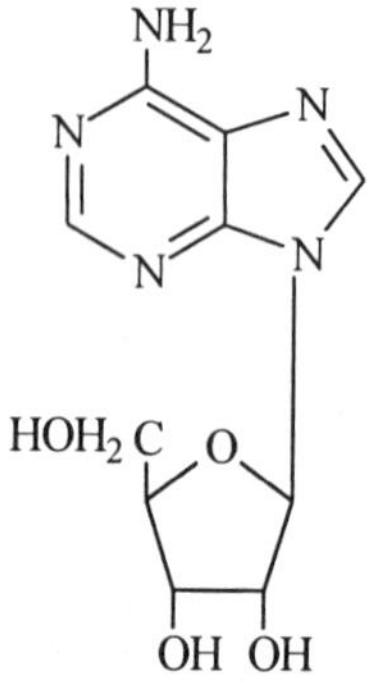

NFB-4

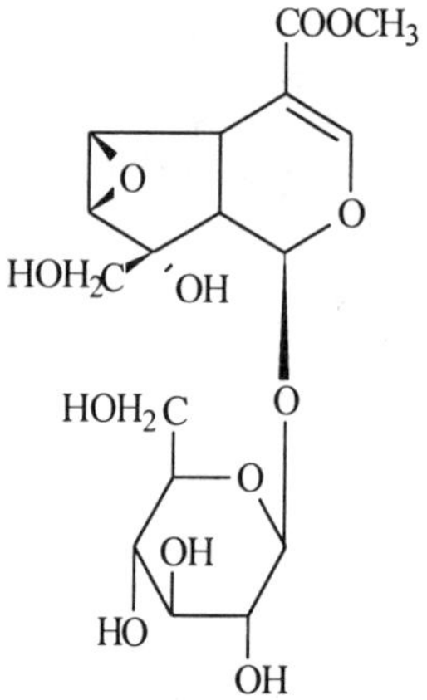

NFB-8

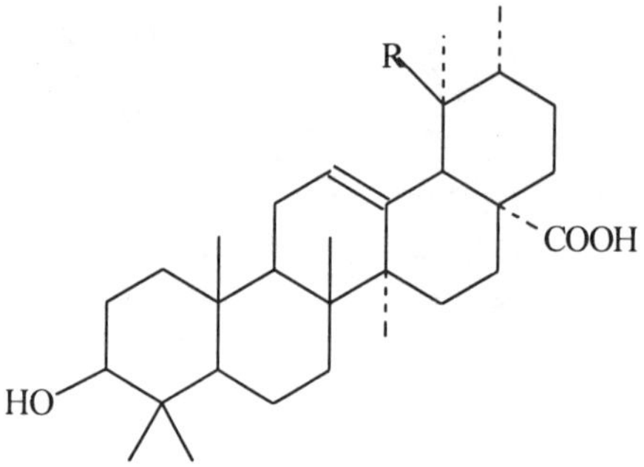

NFE-1: R=H
NFE-2: R=OH

Figure 9. Known compounds identified from the Indian noni fruits.

Analysis of Indian Noni Leaves

As shown in Figure 10, the dried noni leaves (5 Kg) were extracted with 95% ethanol (4 L) at 50°C for 1 day. The extract was concentrated to dryness under reduced pressure, and the residue was suspended in water (500 mL) and partitioned successively with hexane (3 × 500 mL), ethyl acetate (3 × 500 mL), and *n*-butanol (3 × 500 mL). The butanol fraction was subjected to a Diaion HP-

20 column, eluted with water-ethanol (water, 30% ethanol, 70% ethanol, 95% ethanol) solvent system. Four subfractions were obtained. From the ethyl acetate fraction, we obtained 1 compound. From the 30% ethanol fraction, 14 compounds were isolated, 12 of them were identified. Four of them are new iridoids (Figure 11) and the other 8 are known compounds (Figure 12).

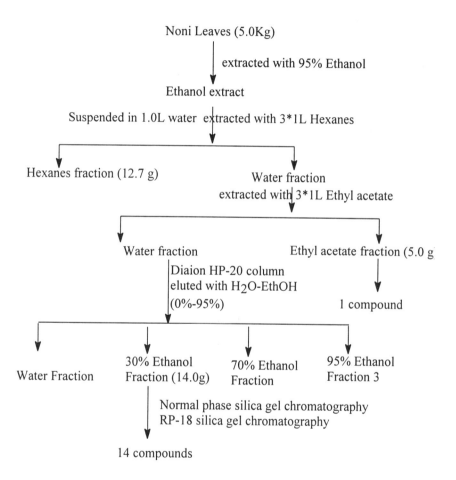

Figure 10. Extraction and isolation procedure for compounds from Indian noni leaves.

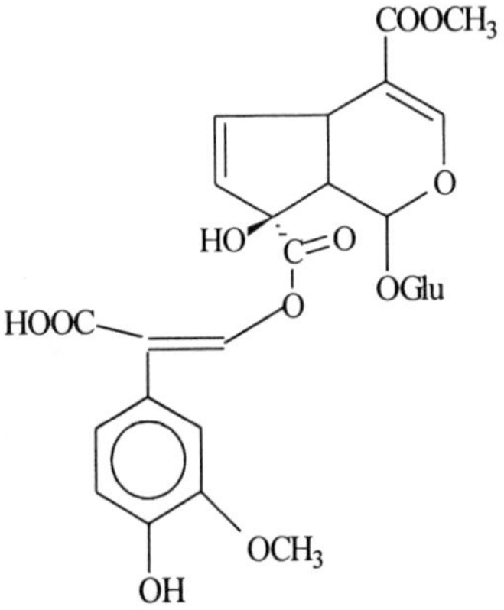

Figure 11. Novel compounds identified from the Indian noni leaves.

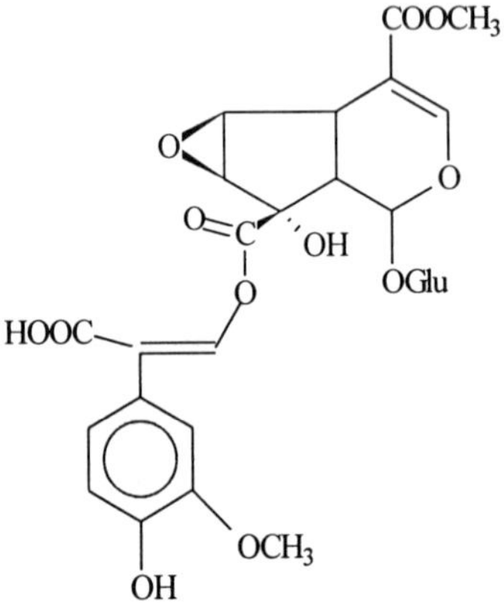

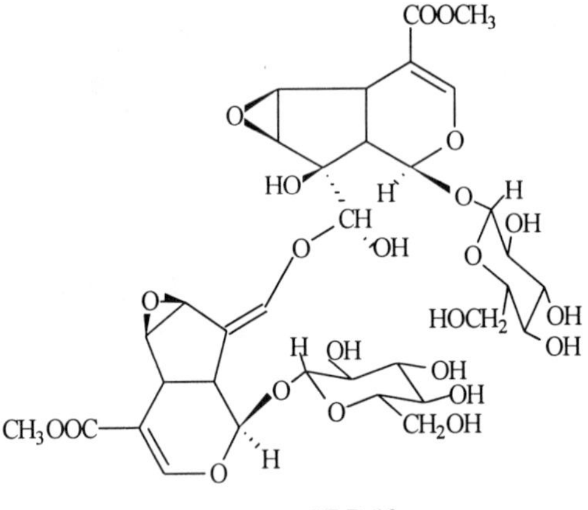

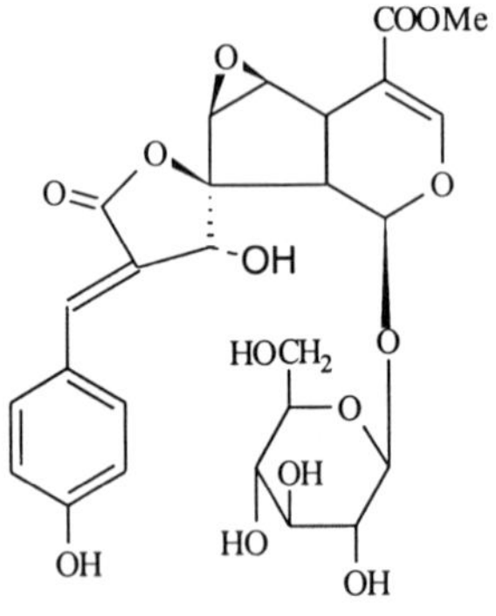

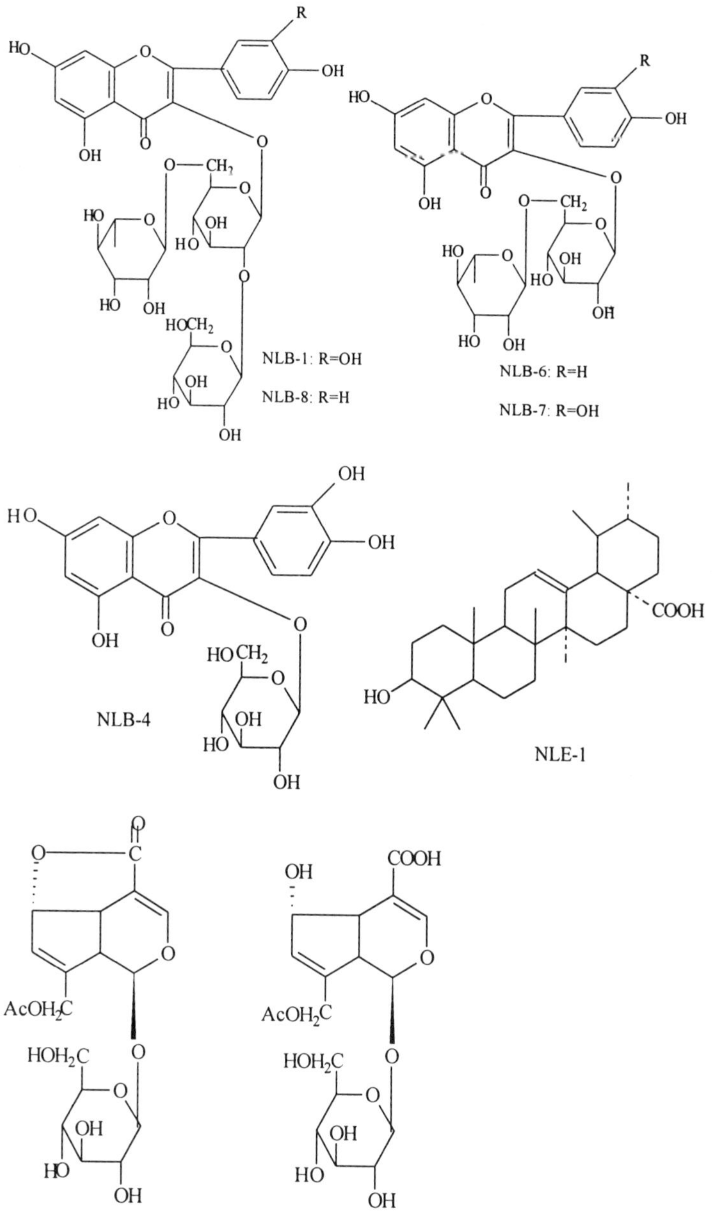

NLB-1: R=OH
NLB-8: R=H

NLB-6: R=H
NLB-7: R=OH

NLB-4

NLE-1

NLB-5: Asperuloside

NLB-11

Figure 12. Known compounds identified from the Indian noni leaves.

Antitumor Promoting Effect of the Components of Noni

It is well known that UVB irradiation acts both as a tumor initiator and tumor promoter, playing a major role in the development of human skin cancer. Because transactivation of AP-1 plays a key role in tumor promotion (*18*), in the present study we investigated the effect of some pure compounds isolated from noni on UV-induced AP-1 activity in both cell cultures and AP-1-luciferase

On the basis of the importance of AP-1 activity in tumor promotion and progession, we tested the inhibitory effect of the UVB-induced AP-1 activity of four pure compounds, **NB-10** isolated from Hawaiian noni fruits, **NLB-2, NLB-3** isolated from Indian noni leaves, and **NB-11** isolated from both the Hawaiian noni fruits and the Indian noni fruits and leaves, which also was the major compound of Indian noni fruits. We also tested the UVB-induced phosphorylation of JNKs, Erks and P-38 activity of **NB-10** and **NB-11**. Figures 13 and 15 showed that all of these four compounds suppressed UVB-induced AP-1 activity. However, **NB-11** and **NLB-3** showed stronger inhibitory effect than NLB-2 and NB-10 in this case. The IC$_{50}$ of **NLB-2** and **NLB-3** are 69.6 and 29.0 μM respectively. Figure 14 showed that **NB-11** but not **NB-10** significantly inhibited the phosphorylation of Erks and P-38 induced by UVB at the concentration of 100 μM. But both of them didn't inhibit the phosphorylation of JNKs induced by UVB.

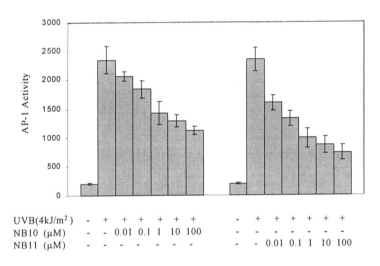

Figure 13. Inhibitory effect of NB-10 and NB-11 on UVB-induced AP-1 activity.

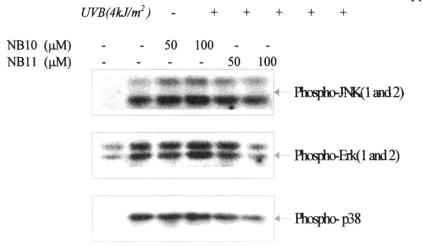

Figure 14. Inhibitory effect of NB-10 and NB-11 on the phosphorylation of JNKs, Erks and p38 induced by UVB.

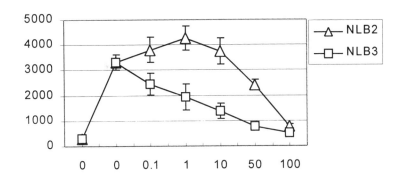

Figure 15. Inhibitory effect of NLB-2 and NLB-3 on UVB-induced AP-1 activity.

Conclusion

In this research, we systematically studied the chemical components of Hawaiian noni fruits, Indian noni fruits and leaves. By the combination of solvent extraction, partition and modern chromatograph methods, including normal phase silica gel column chromatography, reverse phase silica gel column chromatography, prepared thin layer chromatography, sephadex LH-20 column chromatography and Diaion HP-20 column chromatography, a total of about 30 compounds were isolated from noni. Ten of them are new compounds.

According to our research, we find that the constituents of the Hawaiian noni fruits are very different with those of the Indian noni fruits. This indicates that the constituents of the plant can be markedly influenced by geographical conditions.

Studies on the atitumor promoting activity have been carried out with four isolated pure compounds. All of them suppressed UVB-induced AP-1 activity. However, **NB-11** and **NLB-3** showed stronger inhibitory effect than **NLB-2** and **NB-10**. In addition, **NB-11** but not **NB-10** significantly inhibited the phosphorylation of Erks and p-38 induced by UVB at the concentration of 100 μM. But both of them did not inhibit the phosphorylation of JNKs induced by UVB. Further investigations on the prevention of disease and improvement of human health by noni and/or its bioactive components are required.

References

1. Levand, O.; Larson, H. *Plant Med.* **1979**, *36*, 186-187.
2. Hirazumi, A.; Furusawa, E. *Proc.West Pharmacol. Soc.* **1994**, *37*, 145-146.
3. Hirazumi, A.; Furusawa, E. *Proc. West Pharmacol. Soc.* **1996**, *39*, 25-27.
4. Daulatabad, C.D.; Mulla, G.M. *J. Oil Technol. Assoc.* **1989**, *21*, 26-27.
5. Rusia, K.; Srivastava, S.k. *Curr. Sci.* **1989**, *58*, 249-251.
6. Jain, R.K.; Ravindra, K. *Proc. Natl. Sci. India,* Sect. A **1992**, *62*, 11-13.
7. Srivastava, M.; Singh, J. *Int. J. Pharmacogn.* **1993**, 182-184.
8. Singh, J.; Tiwari, R.D. *J. India Chem. Soc.* **1976**, *52*, 424-425.
9. Tiwari, R.D.; Singh, J. *J. India Chem. Soc.* **1977**, *54*, 429-430.
10. Ahmad, V.U.; Bano, M. *J. Chem. Soc. Pak.* **1980**, *2*, 71-73.
11. Farine, J.P.; Legal, L. *Phytochemistry* **1996**, *41*, 433-438.
12. Hirazumi, A.; Furusawa, E. *Phytotherapy Research* **1999**, *13*, 380-387.
13. Hiramatsu, T.; Imoto, M. *Cancer letters*, **1993**, *73*, 161-166.
14. Hiwasa, T.; Arase, Y. *FEBS Letters*, **1999**, *444*, 173-176.
15. Younos, C.; Rolland, A. *Planta Med.* **1990**, *56*, 430-434.
16. Wang, M.; Kikuzaki, H.; Csiszar, K.; Boyd, C.D.; Maunakea, A.; Fong, S.F.T.; Ghai, G.; Rosen, R.T.; Nakatani, N.; Ho, C.-T. *J. Agric. Food Chem,* **1999**, *47*, 4880-4882.
17. Wang, M.; Kikuzaki, H.; Jin, Y.; Nakatani, N.; Zhu, N.; Csiszar, K.; Boyd, C.D.; Rosen, R.T.; Ghai, G.; Ho, C.-T. *J. Nat.Prod.* **2000**, *63*, 1182-1183.
18. Huang, C.S.; Ma, W.Y. *J. Biol. Chem.* **1997**, *272*, 26325-26331

Chapter 11

Analysis and Standardization of Cranberry Products

David G. Cunningham, Sarah Vannozzi, Elizabeth O'Shea, and Richard Turk

Ocean Spray Cranberries, Inc., One Ocean Spray Drive, Lakeville, MA 02349

Folklore has long supported the role of cranberry juice in maintaining urinary tract health. Now, a significant body of scientific evidence supports this cranberry benefit. Recently, bioassay directed fractionation studies have identified cranberry proanthocyanidins as the compounds responsible for preventing the adhesion of certain *E. Coli* to the uroepithelial cells. Prior to this discovery, an established analytical method for determining the percentage of cranberry juice in a product has been used to standardize products to ensure the contained a minimum and consistent amount of cranberry. Now, with this evidence for the active component in urinary tract health, valid analytical methods are needed to assure product claims regarding proanthocyanidin content. The analytical methods and the issues faced in validating methods for cranberry proanthocyanidin quantification will be reviewed.

Folklore for a long time has alluded to the urinary tract health (UTH) benefits of cranberry juice. UTH refers to the healthiness of the urethra, bladder, kidneys and prostate, parts of the urinary tract that are frequently subject to infections (UTIs) caused by pathogenic bacteria, primarily *Escherichia coli* (*E. coli*) (*1*). Drinking cranberry juice was originally believed to decrease the risk and help alleviate the symptoms of these infections by acidification of the urine (*2*). More recent research has instead shown that cranberry juice actually inhibits the adherence of certain types of *E. coli* to the uroepithelial cells lining the urinary tract (*3*).

Cranberry juice was clinically proven to benefit UTH by Avorn (*4*), based on the results of a clinical trial conducted with a population of 153 elderly women. Using a randomized, double-blind, placebo-controlled design, Avorn assigned the women to drink 10 ounces per day, for six months, of either a low-calorie cranberry juice cocktail, or a placebo drink which contained no cranberry juice. The results of the study showed that cranberry juice significantly reduced bacteria and white blood cells, two indicators of UTI, in the urine of the study group. Compared to the control group, those drinking cranberry juice were half as likely to have high bacteria and white blood cell counts, or quarter as likely to continue exhibiting high bacteria and white blood cell counts, in their urine. Further, Avorn confirmed that the cranberry juice cocktail used in the study inhibited *E. coli* adhesion while the placebo drink did not, based on the in-vitro testing of both for bacterial anti-adhesion bioactivity. Finally, Avorn also found that compared to the control group, the urine was less acidic for those drinking cranberry juice, refuting the earlier hypothesis of the UTH benefit being due to urinary acidification.

Cranberry Juice Standardization

Even though the scientific evidence now began to support the long-thought benefit of drinking cranberry juice in maintaining UTH, more research was needed to identify the active component(s) in cranberries responsible for the anti-adhesion bioactivity. Therefore, it was not possible to directly standardize cranberry products for the UTH benefit other than by the anti-adhesion bioassay, which is logistically impractical. However, by measuring the amount of quinic acid in cranberry juice and product formulation, an in-direct standardization of cranberry products for an implied UTH benefit is potentially achievable.

Coppola developed a reverse phase C-18 HPLC/UV method for measuring quinic and the other organic acids in a juice or beverage, primarily as a means to detect cranberry juice adulteration (*5-7*). Quinic acid is an organic acid relatively unique to cranberry in its presence, amount and ratio to other organic acids. The amount of quinic acid measured is compared to the average quinic acid value determined for 7.5-degree brix, single strength cranberry juice, to determine the amount of cranberry juice present. The Association of Official Analytical Chemists (AOAC) accepted this analytical method in 1989 as method 986.13, 'Quinic, Malic and Citric Acids in Cranberry Juice Cocktail and Apple Juice'. Further, in 1985 the USDA supported the standardization of cranberry juice cocktail based on quinic acid content, by establishing a commercial item description (CID) for this product specifying that it must contain a minimum of 25% cranberry juice and 0.26% quinic acid.

Active Component Discovery

After the publication of the Avorn clinical study, research focused on identifying the active component(s) responsible for the bacterial anti-adhesion believed associated with the UTH benefit of cranberries. Howell et al. (8) and Foo et al. (9,10) reported using bioassay directed solid phase extraction (SPE) to isolate and identify the proanthocyanidins (PACs), or condensed tannins, as the active components in cranberries. This fraction was found to have specific in-vitro activity against the adherence of p-fimbriated type E. coli to uroepithelial cells, of the type lining the walls of the urinary tract. In general, the PACs are also being studied for other potential health benefits, such as anticancer (11) and cardiovascular (12).

The PACs Howell et al. and Foo et al. identified in cranberry are of a subclass called the procyanidins, polymers of catechin and epi-catechin that are linked together by B-type (4-6 or 4-8) and A-type (4-8, 2-O-7) interflavan bonds (Figure 1)(8-10). The PACs identified in cranberry, associated with the anti-adhesion bioactivity, will be referred to as UTH-PACs. So far, it appears that UTH-PACs are trimers or larger, containing at least one A-type interflavan bond. In addition, our work indicates that cranberry may have PAC oligomers and polymers up to at least dodecamers, and that may also contain gallocatechin and/or epigallocatechin.

Active Component Quantitation

With the identification of the UTH-PACs as the components in cranberries responsible for the anti-adhesion bioactivity, our interest was to develop a routine, reliably method for their quantification. The ability to quantify UTH-PACs will provide another way, in addition to juice color and quinic acid content, to standardize and manage the overall quality of cranberry products. Howell's bioassay directed SPE and isolation of the anti-adhesion bioactive UTH-PAC fraction of cranberries provided the primary reference method for their selective gravimetric quantitation. In addition we reviewed and evaluated a number of other methods reported on in the literature for quantifying PAC-type compounds, to the Howell method. The type of methods investigated included another gravimetric method, as well as colorimetric, chromatographic and mass spectrometric techniques.

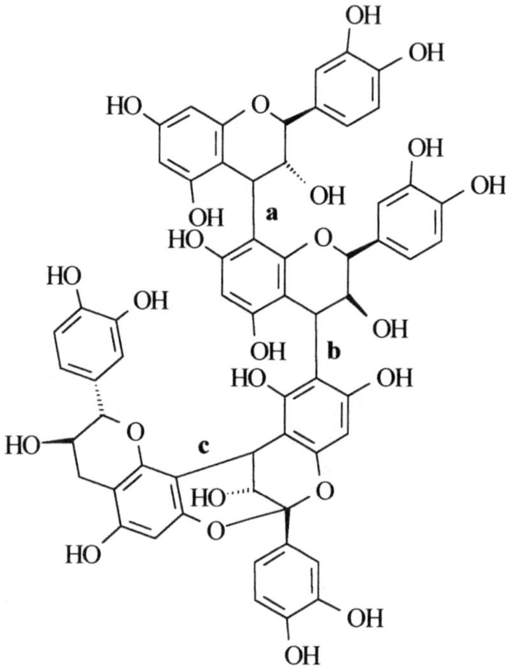

Figure 1. Procyanidin tetramer with a 4β-8 (a) and a 4β-6 (b) B-Type interflavan bond and a 4β-8, 2β-O-7 (c) A-Type interflavan bond.

Sample Preparation

Initial sample preparation was a common requirement for all the methods investigated. While liquid samples (e.g.; juice) could be readily analyzed, a PAC liquid extract was needed for solid samples (e.g.; fruit). A liquid-solid extraction procedure using 80% acetone was validated for fruit and other solid samples. Solid samples are ground to a liquid slurry in a stainless steel Waring blender container with liquid nitrogen, resulting in a fine powder after nitrogen evaporation, that is kept frozen. A weighed sample of the powder is extracted with 80% acetone, vortexed, sonicated, centrifuged and filtered. The extraction procedure and solvent were determined through personal communication, literature (*13-15*) and an experiment to show the effectiveness of several organic solvent choices. The organic solvents for the experiment, all at 80% in water, were chosen for their dielectric constants as a measure of their polarity, and ease of removal from the aqueous phase. The 80% acetone solution was ultimately chosen for its high recovery, tight standard deviation and low boiling point, a result supported by what had been found in the literature cited.

Gravimetric Methods

Howell Method

Howell used reverse phase (C18) followed by adsorption chromatography (Sephadex LH-20) to fractionate and isolate the UTH-PACs (*16*). An aqueous sample/extract is loaded onto a C18 column, where water, then 15% methanol are used to elute off sugars and acids, and acidified methanol is used to recover the sample. The sample is dried, reconstituted in 50% ethanol and loaded onto a Sephadex LH-20 column. The flavonols and anthocyanins are eluted off with 50% ethanol and 70% acetone is used to recover the UTH-PAC fraction. The UTH-PAC elution is centrifugally evaporated or lyophilizied to produce a dried material that is weighed to quantify the UTH-PACs. This is a non-destructive analysis, where the dried UTH-PAC extract can be used for further analysis or as a source material for other experiments.

Sample throughput is a major problem using this method for any laboratory trying to analyze a large number samples. Further, though the sample size analyzed can be adjusted, the method's overall sensitivity is limited by the ability to weigh milligram or smaller quantities of UTH-PAC extract in the relatively heavy tubes or flasks used to dry the material. These problems translate into the need for larger sample sizes, requiring larger scale separations and resulting in higher costs for associated materials and technician time. One other method drawback is that it is not qualitative; it gives no information about individual UTH-PAC composition. In addition, there is potential for measurement error arising from the transfer steps required prior to the final weighing.

Trivalent Ytterbium Precipitation Method

Another gravimetric method that was reported on by Reed (*17*) involves the selective precipitation of PACs with trivalent ytterbium to recover this fraction. Reed adapted his procedure for cranberries based on the work of Giner-Chavez (*18*) where Sephadex LH-20 is used to remove organic acids and other non-tannin phenolics from the sample, which would otherwise be precipitated by the trivalent ytterbium, instead of, or in addition to the PACs. An aqueous sample/extract is loaded onto the column; water is used to elute off acids and 80% acetone is used to recover the PAC fraction. The acetone is removed by evaporation, triethanolamine added to raise the pH and 0.1 M ytterbium acetate added to precipitate the PACs. The precipitate is washed and recovered by centrifugation, oven dried, weighed hot, ashed and re-weighed hot. The PAC content is determined by the difference in sample weight before and after ashing. The quantitation of the sample is destructive. If PAC material is to be recovered for further analysis or use, the sample precipitate must be quantitatively split. One part is dried and ashed, while the other part is treated using ion exchange to recover PACs from the trivalent ytterbium. Compared to the Howell method, this method also has problems with of sample throughput, sensitivity and lack of qualitative information. In addition, an extra step is needed to recover PAC material for further analysis or use.

Colorimetric Methods

Vanillin Assay

Colorimetric methods were investigated since they had the potential to be more sensitive than the gravimetric methods, and depending on the chemistry of the colorimetric reaction, could provide an additional degree of selectivity. The vanillin hydrochloric acid method was considered, but determined not to be well suited for use with cranberries. Its been reported that it may not be as specific as needed (*19*) and that it may be prone to interference from vitamin C (*20*), which is naturally present in cranberries, and acetone, the preferred elution solvent (*21*). In addition, residual traces of the anthocyanin pigments of cranberries, which have an absorption maximum at 520 nm, could interfere at the method's 500 nm measurement wavelength.

DMAC Assay

The aldehyde condensation of 4-dimethylamino-cinnimaldehyde (DMAC) was a colorimetric method reported to work for measuring residual tannin in beer after chill proofing (*19*). DMAC reacts with the terminal monomer of a polymeric PAC, at the eight-carbon position of the flavonoid A-ring. The use of DMAC appeared to be better suited for cranberry, since the color developed by the reaction is read at 640 nm. Combined with the LH-20 Sephadex SPE procedure used in the trivalent ytterbium precipitation method, the method could also be made selective for UTH-PACs. Based on this preliminary review, the DMAC method was developed, optimized and validated in-house for use with cranberry samples. The DMAC method has resulted in a quick and reliable procedure for fairly inexpensive, high throughput work.

An aqueous sample/extract is loaded onto the Sephadex LH-20 column. Water, then 50% ethanol is used to elute off the anthocyanins, flavonols and non-UTH-PACs that could potentially react with DMAC. The PAC fraction is recovered from the LH-20 using 80% acetone. If needed, the concentration can be adjusted by dilution to ensure the absorption reading will be on scale. The PACs in the sample, on reaction with DMAC, will result in a blue/green colored solution that can be read at 640 nm, an UV region with little or no response for cranberries.

The DMAC method was standardized against the Howell method using a select lot of Ocean Spray 90MX powder; a spray dried cranberry juice product, which was chosen as a standard UTH-PAC source material. The average UTH-PAC content of the 90MX powder was determined using the Howell method and this value used to establish a response factor for the DMAC method. This 90MX powder is analyzed with each batch of samples and the result control charted to ensure the method is not out of calibration. In addition, this 90MX powder was analyzed as an unknown by both methods (n=10) to determine the precision of each method, resulting in a 4.6% RSD for the DMAC method and 7.5% RSD for the Howell method. Finally, 22 different samples were each analyzed by both methods and it was found that the average difference between the Howell and DMAC methods for the samples was 0.3%. A paired comparison t-test of these results showed no significant difference at the 99% confidence interval.

In comparison to the gravimetric methods, the DMAC colorimetric method developed is faster (one-day versus two-three days) and more sensitive. With improved sensitivity, sample amounts can be adjusted accordingly, resulting in a scaled-down Sephadex LH-20 SPE step, which results in a smaller apparatus setup and the ability to physically analyze an increased number of samples. The cost of the analysis is greatly reduced over the Howell method with savings both in materials and technician time, making it our lab's method of choice for routine PAC analyses. Other colorimetric methods used for the more general, non-specific analysis of phenolics and tannins are the Folin Ciocalteu (22), the n-butanol hydrochloride (23) and the Prussian blue (24) procedures.

Chromatographic Methods

Chromatographic methods are capable of both qualitatively characterizing the UTH-PAC composition of cranberry samples as well as quantifying the individual PACs. Recently, Hammerstone (25) published a normal phase liquid chromatographic (NP-HPLC) method for separating PACs in cocoa. The method uses a 25-cm silica column and a ternary or binary mobile phase gradient of methylene chloride, methanol and water:acetic acid (1:1), with detection by atmospheric pressure ionization electrospray (API/ES) mass spectrometry, in scan or selected ion monitoring (SIM) mode. Detection by UV absorbance (280 nm) and fluorescence (276 nm excitation/316 nm emission) (26) were also used for comparison. As mentioned, the method was initially developed for the analysis of PACs in cocoa, but has also been used for the comparative analysis of the PACs in apples, peanuts and almonds, grape seeds and juice, wine and tea (26). To quantify PACs, Adamson separated and collected individual PAC peaks from cocoa for calibration standards, since there are only a few PAC standards (mostly monomers and dimers) commercially available (27).

Using Hammerstone's NP-HPLC method for the qualitative analysis of cranberry extracts shows a much more complex PAC composition than seen for cocoa. Cocoa appears to have PACs that are mostly homogeneous (epi-catechin) with all B-type interflavan bonds. Cranberry, as already noted, have PACs that are more heterogeneous (epi-catechin and epi-gallocatechin) with both B-type and A-type interflavan bonds. So where cocoa may have a single peak for a PAC with a specific degree of polymerization (DP), cranberry will potentially have a number of peaks due to a variety of possible isomers, and an increasingly greater number, the greater the PAC's DP. The result is a chromatogram with fairly well resolved peaks for PACs up to tetramers, but with a large, unresolved rise in the baseline for PACs of higher DP. Unfortunately, it's this region that is of the most importance to cranberry. Figure 2 compares the NP-HPLC analysis of a cranberry extract prepared using the Howell SPE method to one prepared using a modification of the SPE procedure described by Hammerstone for cranberry and blueberry (28). The NP-HPLC analysis shows no PAC monomers and relatively small PAC dimer and trimer peaks in the Howell extract, whereas the Hammerstone extract has

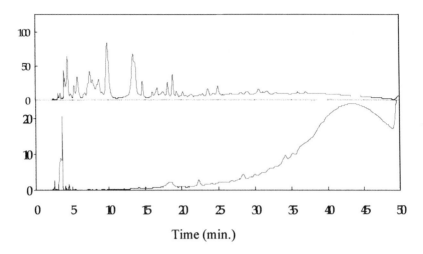

Figure 2. HPLC/DAD chromatograms showing the Hammerstone SPE extract (top) and the Howell SPE extract (bottom).

large PAC monomer, dimer and trimer peaks. The Howell extract consists mostly of the late-eluting unresolved complex of peaks, where the higher DP PACs are expected to elute, while the Hammerstone extract has only a few small peaks in this region. This makes it difficult to both qualitatively assess and quantify the UTH-PACs in cranberry using the NP-HPLC method. Figure 2 also points out the need to consider how sample extracts are prepared before comparing PAC results.

Using the NP-HPLC method to quantify the PAC oligomers in cranberry would be extremely difficult following the procedure Adamson described to obtain a set of calibration standards, due to the complexity seen in cranberry. Since the diversity of PACs in cranberry would make the task of recovering and purifying each PAC monumental, an alternative is to establish a method calibration using a relative standard, such as epi-catechin. Each peak in the chromatogram would have its amount reported based on the detector's response to epi-catechin. When using a relative standard to quantify the PACs, differences in compound dependent detector responses, such as UV molar absorptivity or ionization efficiency and transfer kinetics in mass spectrometry, can impact absolute accuracy. Another approach that would relatively quantify the PACs in cranberry is to use a standard sample, such as 90MX powder, in the same manner used to standardize the DMAC method. In this case, the total area of the peaks of a 90MX sample is compared to the total area of the peaks of an unknown cranberry sample.

Mass Spectrometric Methods

Mass spectrometry provides a powerful tool for both characterizing and quantifying the PAC composition of cranberries. Mass spectrometry, when used in conjunction with liquid chromatography provides a "two dimensional" (time and mass) analysis, as described above, or when used alone can provide a fast, "one dimensional" (mass) analysis. When the mass spectrometer (MS) is used, as a detector for the NP-HPLC method, the instrument can be set up to scan over a number of narrow, discrete mass ranges, centered on the molecular ions of individual PAC polymers. This technique minimizes the scan time, increasing the number of scans that can be made and thereby maximizing sensitivity. Figure 3 shows extracted ion chromatograms for six scanned mass ranges, corresponding to PAC monomers to hexamers. It should be noted that at the retention time seen for a PAC tetramer, there is a response in both the tetramer ion (1152 m/z) and dimer ion (575 m/z) chromatograms. The dimer ion response is due to the presence of a doubly charged PAC tetramer ion, an ionization effect that can be seen when using API/ES mass spectrometry in the analysis of large molecules. In fact, the ability of PACs molecules with a high DP to pick up an additional charge makes them visible below 1800 m/z, the upper limit of our quadrupole LC/MS/MS instrument. Therefore, the presence of specific PAC polymers can be confirmed based on the retention time, and both the molecular ion mass, and the existence of multiply charged ion masses. Extracted ion chromatograms can also help differentiate PACs based on the type and number of interflavan bonds present. Figure 4 shows superimposed ion chromatograms for a trimer with all B-type bonds (MW 865.8) and a trimer with one A-type bond (MW 863.8). Comparing these extracted ion chromatograms, a small offset in retention times can be observed. The relative response of the two peaks also indicates that there may be more of the PAC trimer with one A-type interflavan bond in this cranberry sample. As previously mentioned, absolute quantitation of these individual PACs is dependent on the availability of standards, or if using a single standard, an understanding of the factors affecting relative mass spectral response.

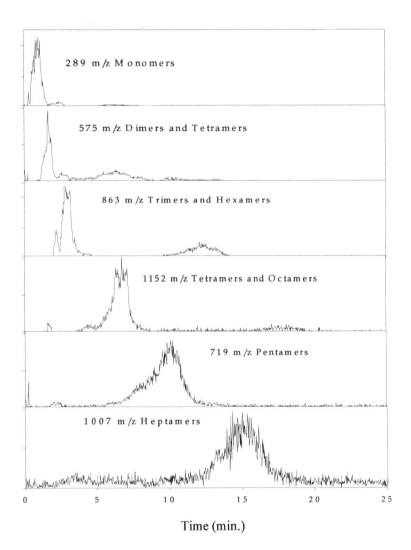

289 m/z Monomers

575 m/z Dimers and Tetramers

863 m/z Trimers and Hexamers

1152 m/z Tetramers and Octamers

719 m/z Pentamers

1007 m/z Heptamers

| 0 | 5 | 10 | 15 | 20 | 25 |

Time (min.)

Figure 3. Extracted ion chromatograms of cranberry PAC extract analyzed on a 5-cm silica column showing monomers through octamers.

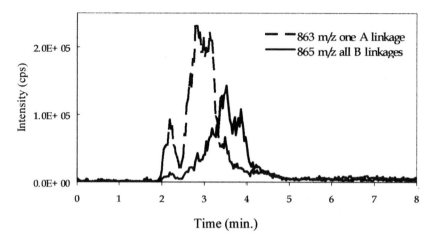

Figure 4. Extracted ion chromatograms of cranberry PAC extract analyzed on a 5-cm silica column showing a trimer with one A-type bond (m/z 863) and a trimer with all B-type bonds (m/z 865).

As shown in Figure 2, most of the UTH-PACs elute later in the NP-HPLC method as an unresolved complex of peaks. One reason for this unresolved complex may be band broadening of these late eluting peaks. Band broadening also results in a loss of sensitivity, as the concentration of a component at the detector is diluted out over the time it takes the peak to elute off the column. To be able to better detect the late eluting PACs, more sample can be injected, but this solution is limited by column overloading effects. However, by removing the liquid chromatography separation step, an unlimited amount of sample can be directly introduced into the MS, allowing the UTH-PACs to be more easily seen. Krueger used matrix assisted laser desorption ionization/time of flight (MALDI/TOF) mass spectrometry to analyze PACs recovered from a trivalent ytterbium precipitate of a grape seed extract (*29*). The mass spectrum obtained provided a qualitative profile of the PAC composition of the sample extract, with Krueger observing a series of PAC polymers in their sodium adduct form. This work showed that MALDI/TOF mass spectrometry is a useful technique for the analysis of PACs, capable of analyzing their wide, high molecular weight range, with very good mass resolution.

More recently, using direct infusion of a PAC extract into an API/ES MS source, our lab has demonstrated the same potential to qualitatively describe the PAC composition of a sample extract, as seen in Krueger's MALDI/TOF work. Figure 5, shows the mass spectrum obtained from the infusion analysis of a cranberry PAC extract. Since it appears that PACs form stable molecular ions under API/ES conditions, each ion in this mass spectrum can be associated with a specific PAC. Assignments are made using the calculated mass to charge (m/z) ratio for a PAC, based on its DP, monomeric composition, number of A-type interflavan bonds present and number of API/ES charges. For example, a PAC undecamer, with all epi-catechin monomers, one A-type interflavan bond and three charges, will have a mass to charge ratio of 1056, an ion response seen in Figure 5 that can not be achieved by any other reasonable combination.

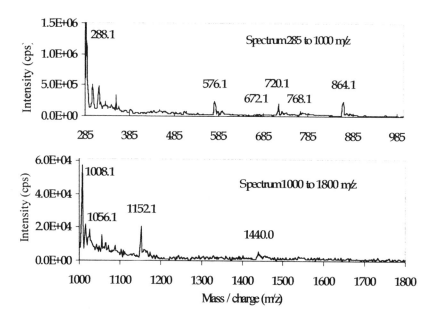

Figure 5. Full scan spectrum (100 scans) of cranberry PAC extract obtained by infusion into API/ES/MS source.

Summary

Table I provides a summary of the methods reviewed in this report, focusing on how fast, sensitive, accurate and reliable each method is. Based on the outcome of this work, the DMAC method most appropriately meets our needs for routinely quantifying cranberry UTH-PACs. This method is fast, taking no more than one day to run a batch of samples. The cost of supplies for this method has been reduced over the Howell method. The method has a relative standard deviation of 4.6 percent, and when analyzing 12 samples of Ocean Spray Cranberry Juice Cocktail ® obtained from four different bottling locations, resulted in an average UTH-PAC content of 30 mg/8 fl. oz., with a relative standard deviation of 14.3 percent. Finally, when compared to the Howell method for a range of different samples, there is no significant difference in the results for this method, at the 99 percent confidence interval. The advantages and disadvantages of each of the methods reviewed show some of them to be better suited for characterizing than quantifying PACs. Many of these methods will be used to further research, which PACs are present in cranberries, and to characterize those PACs found.

Table I. Summary of UTH-PAC Analysis Methods

Method	Howell	Yb precipitate	DMAC	HPLC	LC/MS	Direct ES/MS
Time	4 days	2 days	1 day	Prep + ½ day	Prep + ½ day	Prep + 20 min.
	5 mg in sample	5 mg in sample	15 ug/mL	--	--	NA
Qualitative	No	No	No	Yes	Yes	Yes
Quantitative						
-Precision	6.9 %RSD	--	5.5 %RSD	--	--	NA
-Accuracy	Reference method	--	N.S.D. @ 99%	Poor	Poor	NA

References

1. Sobell, J.D. *Med. Clin. North Amer.* **1991**, *75*, 253-273.
2. Blatherwick, N.R.; Long, M.L. *J. Bio. Chem.* **1923**, *57*, 815-818.
3. Ofek, I.; Goldhar, J.; Zafriri, D.; Lis, H.; Adar, R.; Sharon, N. *N. Engl. J. Med.* **1991**, *324*, 1599.

4. Avorn, J.; Monane, M.; Gurwitz, J.H.; Glynn, R.J.; Choodnovskiy, I.; Lipsitz, L.A. *J. Amer. Med. Assoc.* **1994**, *271*, 751-754.
5. Coppola, E.D.; Conrad, E.C.; Cotter, R. *J. Assoc. Off. Anal. Chem.* **1978**, *61*, 1490-1492.
6. Coppola, E.D. *Food Technology* **1984**, *38*, 88-91.
7. Coppola, E.D.; Starr, M.S. *J. Assoc. Off. Anal. Chem.* **1986**, *69*, 594-597.
8. Howell, A.B.; Vorsa, N.; Marderosian, A.D.; Foo, L.Y. *N. Engl. J. Med.* **1998**, *339*, 1085.
9. Foo, L.Y.; Lu, Y.; Howell, A.B.; Vorsa, N. *Phytochemistry* **2000**, *54*, 173-181.
10. Foo, L.Y.; Lu, Y.; Howell, A.B.; Vorsa, N. *J. Nat. Prod.* **2000**, *63*, 1225-1228.
11. Guthrie, N. *The FASEB Journal* **2000**, *14*, A771.
12. Krueger, C.G.; Porter, M.L.; Wiebe, D.A.; Cunningham, D.G.; Reed, J.D. *Polyphenols Communications 2000, Volume 2*, XXth International Conference on Polyphenols, Freising-Weihenstephan, Germany, September 11-15, 2000; Technische Universität München, 2000; p 447.
13. Kiehne, A.; Lakenbrink, C.; Engelhardt, U.H. *Z Lebensm Unters Forsch A* **1997**, *205*, 153-157.
14. Karchesy, J.J.; Hemingway, R.W. *J. Agric. Food Chem.* **1986**, *34*, 966-970.
15. Whittle, N.; Eldridge, H.; Bartley, J. *J. Inst. Brew.* **1999**, *105*, 89-99.
16. Howell, A.B. Rutgers University, Chatsworth, NJ. Unpublished work, 1999.
17. Reed, J.D.; Horvath, P.J.; Allen, M.S.; Van Soest, P.J. *J. Sci. Food Agric.* **1985**, *36*, 255-261.
18. Giner-Chavez, B.I.; Van Soest, P.J.; Robertson, J.B.; Lascano, C.; Reed, J.D.; Pell, A.N. *J. Sci. Food Agric.* **1997**, *74*, 359-368.
19. McMurrough, I.; McDowell, J. *Anal. Biochem.* **1978**, *91*, 92-100.
20. Sun, B.; Ricardo-da-Silva, J.M.; Spranger, I. *J. Agric. Food Chem.* **1998**, *46*, 4267-4274.
21. Makkar, H.P.S.; Becker, K. *J. Chem. Ecol.* **1993**, *19*, 613-621.
22. Singleton, V.L.; Rossi Jr., J.A. *Am. J. Enol. Vitic.* **1965**, *16*, 144-158.
23. Porter, L.J.; Hrstich, L.N.; Chan, B.G. *Phytochemistry*, **1986**, *25*, 223-230.
24. Price, M.L.; Butler, L.G. *J. Agric. Food Chem.* **1977**, *25*, 1268-1273.
25. Hammerstone, J.F.; Lazarus, S.A.; Mitchell, A.E.; Rucker, R.; Schmitz, H.H. *J. Agric. Food Chem.* **1999**, *47*, 490-496.
26. Lazarus, S.A.; Adamson, G.E.; Hammerstone, J.F.; Schmitz, H.H. *J. Agric. Food Chem.* **1999**, *47*, 3693-3701.

27. Adamson, G.E.; Lazarus, S.A.; Mitchell, A.E.; Prior, R.L.; Cao, G.; Jacobs, P.H.; Kremers, B.G.; Hammerstone, J.F.; Rucker, R.B.; Ritter, K.A.; Schmitz, H.H. *J. Agric. Food Chem.* **1999**, 47, 4184-4188.
28. Hammerstone, J. F.; Lazarus, S. A. M&M Mars, Hackettstown, NJ. Unpublished work, 2000.
29. Krueger, C.G.; Dopke, N.C.; Treichel, P.M.; Folts, J.; Reed, J.D. *J. Agric. Food Chem.* **2000**, 48, 1663-1667.

Chapter 12

Extraction of Antioxidants from Grape Seeds Employing Pressurized Fluids

L.T. Taylor[1] and Miguel Palma[2]

[1]Department of Chemistry, Virginia Institute of Technology and State University, Blacksburg, VA 24061
[2]Department of Analytical Chemistry, University of Cádiz, 11510 Puerto Real, Spain

Abstract

White grape seeds were subjected to sequential supercritical fluid extraction. By increasing the polarity of the supercritical fluid using methanol as a modifier of CO_2, it was possible to fractionate the extracted compounds. Two fractions were obtained; the first, which was obtained with pure CO_2, contained mainly fatty acids, aliphatic aldehydes, and sterols. The second fraciton, obtained with methanol-modified CO_2, had phenolic compounds, mainly catechin, epicatechin, and gallic acid. The fractions were bioassayed. Antimicrobial activities were checked on human pathogens, and a high degree of activity was obtained with the lipophilic fraction.

Wine and wine-derived products are made in many countries around the world. Many chemical compounds must be monitored during the wine making process in order to obtain high quality wine. For example, compounds related to flavor, color, and aroma must be analyzed to increase the organoleptic properties of the wine. Metals and some organic acids must be controlled to avoid degradation processes like precipitation.

Since wine is a liquid, analysis of compounds is easier than in a semi-solid or solid matrix because no extraction nor pre-treatment steps are required. For wine, only pre-concentration steps are necessary for compounds in low concentration. Wine composition, however, is clearly determined by the grape used for obtaining the wine. Since grape is a non-liquid material, extraction steps are needed for compound determination. Extraction methods for grape can be designed to be similar to the extraction process for wine (i.e. pressing the grape). From an analytical point of view, this option is inadvisable due to the fact that it would be a time-consuming process. Therefore, extraction with organic solvents is more widely used. Specific solvents can be used to obtain specific extracts, thus making subsequent analysis steps easier.

Grape has different parts which contribute different compounds to the wine. The chemical composition of pulp, skin, and seeds are quite different. Their physical properties are also different, so different extraction methods have been developed. Besides the grape, there are other solid materials which contribute to the composition of wine such as barrel wood and the fruits used as additives in spirits. These solids are also extracted and the extract composition is of interest for the wine making process.

During the past few years it has been demonstrated that phenolic compounds contribute to both the flavor properties and the pharmacological effects of wine. In this regard, phenols in grape seed are known to significantly contribute to the total phenolic composition of the grape-derived wine. This comes about due to several maceration steps that must be taken in the processing of grapes to make wines. It is at the maceration process that phenols are extracted from the seed. Therefore, the contribution of phenolic compounds from seeds is increased when long processing times are used which is most notable in the production of red wines (1). Phenolic compounds such as catechins and procyanidans from seeds have been shown to affect the bitterness and astringency of wines (2-4). In terms of pharmacological properties, these same phenols can act against _in vitro_ oxidation of low density lipoprotein (5). Other grape-derived phenols have also been suggested to have antiulcer(6) anticarcinogenic (7), antimutagenic (8), and antiviral(9) activities. In each case, the high antioxidant power of phenols is generally believed to account for these activities (10,11). Common methods for the isolation of phenolic compounds from grape seed use organic solvents such as methanol (12-14), ethanol (14,15), and acetone (16). Extraction recoveries have been shown to improve if mixtures of methanol/water and acetone/water are employed (15). Various extraction temperatures have been used with times that range from a few minutes to several hours. Since many of the isolation procedures are performed under heated/lighted aerobic conditions, one may question whether the extract represents the true antioxidant activity of the sample.

Supercritical fluid extraction (SFE) with CO_2-based fluids affords big advantages over more conventional extraction techniques. The absence of both light and air during the extraction process can reduce the incidence of degradation reactions. For example, Tipsrisukond, et al. have reported that supercritical fluid extracts exhibit higher antioxidant power than extracts obtained by classical methods *(17)*.

SFE has been previously applied to grape seeds for the removal of oils *(18)*. The main interest in grape seed oils lies in the level of unsaturated fatty acids, such as linoleic and oleic acids. Grape seed oil also contains high quantities of tannins (e.g. 1000 x higher than found in other seed oils). Pure CO_2 without organic modifier was used in this supercritical extraction. The use of small quantities of organic solvent with supercritical CO_2 (i.e. modifier) should cause extraction of phenols from the seed. Since the addition of modifier to the CO_2 raises the critical temperature, the extraction would be under near critical conditions rather than supercritical conditions. Consequently, we have chosen to refer to the combined group of extractions in this paper as pressurized fluid extractions.

In this paper we describe the optimized extraction conditions for removal of eight polyphenols from white grape seeds using modified CO_2. Extraction variables such as a) supercritical fluid CO_2 density, b) nature of organic modifier, c) modifier percentage, and d) the extraction temperature have been considered. Although the distribution of polyphenols is not homogeneous inside the seed, we nevertheless analyzed the global grape seed composition as opposed to analyzing different parts of the seed. Each extract was kept cold and protected from air and light for no more than 24 hours prior to chromatographic analysis. Since the age of the raw material can modify its phenolic composition, our goal was to develop a method to extract the phenolic compounds from grape seed, not to quantify (in an absolute sense) phenols in these particular grape seeds. In the search for practical applications of extracted compounds, the antimicrobial activity was checked using assorted microorganisms.

Experimental

Sample Preparation

Grape seeds were provided by Synthon, Inc. (Blacksburg, VA). The variety of grapes was Chardonnay. They were cultivated in Washington State and handpicked during the harvest of 1997. Seeds were crushed in a coffee grinder for two minutes, but at 15 second intervals the process was stopped for 15 seconds to avoid heating of the sample. The crushed seeds were stored at room temperature prior to extraction.

Extraction Process

For the polyphenol optimization process, seed extractions in stainless steel vessels (1.0 mL) were performed in duplicate. Approximately 30 mg of crushed grape seeds were used in each extraction. The optimization study varied only four parameters: CO_2 density (0.85 vs. 0.95 g/mL), modifier type (methanol vs. ethanol), percent modifier (10% vs. 40%), and extraction temperature (35° vs. 55°C). The following parameters were common to every extraction: liquid CO_2 flow rate = 1 mL/min, amount of organic modifier used in the static mode = 0.25 mL, static extraction time = 20 min., mass of CO_2 used during dynamic extraction = 20 grams, restrictor temperature = 50°C, solid phase trap temperature = 35°C, trap rinse solvent = methanol (3 mL) at 0.5 mL/min.

A much larger seed sample was used in our pressurized fluid fractionation studies. Extractions were conducted in 8 mL stainless steel vessels. A mixture of 7.5 g of crushed grape seeds (1.5 g) and sand (6.0 g) was employed. The extraction conditions are listed in Table I. Fraction A was produced with 100% CO_2; while, Fraction B was obtained by re-extracting the seed sample using the previously optimized conditions.

Apparatus

An Isco-Suprex PrepMaster, Autoprep 44 (Lincoln, NE) equipped with a SSI 222D HPLC pump was used for the supercritical fluid extraction. The solid trap used on the supercritical fluid extraction was packed with Upchurch-C18 (0.81 g) from Chrom Tech (Apple Valley, MN). Carbon dioxide, with helium headspace, was obtained from Air Products and Chemicals Inc. (Allentown, PA). A GC Hewlett Packard (Wilmington, DE) 5890 Series II gas chromatograph equipped with a Hewlett Packard MSD 5972 and a HPLC Series 1050 from Hewlett Packard (Little Falls, DE) equipped with an autosampler, quaternary pump and an UV-Visible multiwavelength detector were used for analysis of the extracts.

Analysis

GC analysis was done with a DB-5MS column (0.25 mm x 30m, dp=0.25μm) from J&W Scientific (Folsom, CA). The injector temperature was 250°C and the flow of carrier gas (He) was 1 mL/min. HPLC analysis was done using a Luna-C18 column (150 x 2 mm, 5 μm particle size) from Phenomenex (Torrance, CA). The solvents were A (2% acetic acid in water) and B (2% acetic acid in methanol). The gradient flow rate was 0.5 mL/min. The gradient schedule was 100% A for 20 min, then 5% B for 5 min, 10% for 15 min, 30% B for 5 min, 40% B for 3 min and finally 100% for 12 min.

Table I. Optimized Conditions for Grape Seed Extractions*,a

	Fraction A	Fraction B
Pressurized Fluids	CO_2 (100%)	CO_2/MeOH (5:1)(v/v)
Solid Trap	C18	C18
Restrictor Temperature	55C	55C
Pressure	450 atm	450 atm
Extraction Temperature	35C	35C
PFE Static Time	15 min	15 min
Dynamic Fluid Mass	25 g	25 g
Trap Rinsing Solvent	10 mL of Hexane	10 mL of Methanol

*1.5 g of seeds

[a]Reproduced with permission from Journal Agriculture and Food Chemistry, 47, (1999) 5044-5048.

Bioassays

Microbial bioassay utilized human pathogens: Gram-positive and Gram-negative bacteria: *Bacillus cereus* (+), *Staphylococcus aureus* (+), *Staphylococcus coagulans niger* (+), *Citrobacter freundii* (-), *Escherichia cloacae* (-), *Escherichia coli* (-), and the fungus *Asperegillus flavus*, the strain used being an aflatoxin producer, and the fungal phytopathogens: *Botrytis cinerea*, *Cladosporium echinulatum*. *Pencillium griseofulvum* was also used as a genus representing the Pencillia. The Gram + and Gram – bacteria were grown on nutrient agar and the fungi on potato-dextrose agar. Each organism was heavily seeded onto diagnostic sensitivity test agar (DST) in plastic Petri dishes, to ensure a dense lawn. To these were added 4-mm disks (Whatman #3 paper) that had been impregnated with various concentrations of fractions A and B. The fractions were placed on the seeded agar surface. Plates were incubated at 37 C and 18 h. Then the diameter of the inhibited area was measured. All data were statistically analyzed.

Results and Discussion

A fractional factorial experimental design that covers half of the possible experiments was employed. The target value in the experimental design was the area of the combined chromatographic peaks derived from the extract. In other words, eight polyphenolic peaks in the typical extract chromatogram were chosen (noted by * on the chromatogram shown in Figure 1), and their areas were added to obtain a total area. Due to the fact that different amounts of sample were taken for various experiments and the final extract volume varied, the total area was divided by the amount of each sample and by the area of the

172

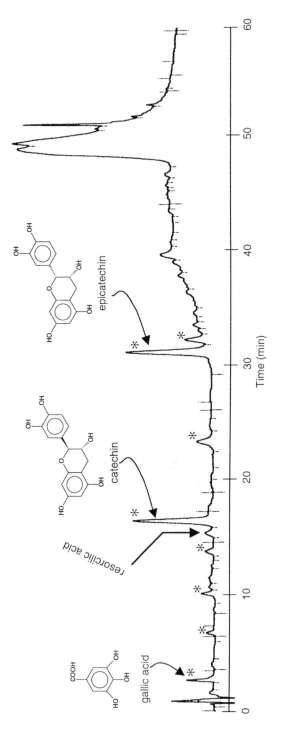

Figure 1. Liquid Chromatogram of polyphenols extract. Reproduced with permission from Journal of Agriculture and Food Chemistry, 47 (1999) 5044-5048.

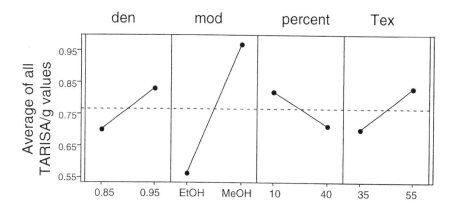

Figure 2. Main effects plot of variables on the average of TARISA(Total Area Relative to Internal Standard Area) per gram of seed. Reproduced with permission from Journal of Chromatography, 849 (1999) 117-124.

internal standard (β-resorcylic acid) used. In this way, the total phenol area per gram of sample and per unit area of internal standard was obtained.

A graphical analysis of the data from the experimental design was made. Figure 2 reveals that the most important variable was the identity of the organic modifier. For example, approximately twice as much polyphenol was removed with methanol as opposed to ethanol. CO_2 density, percentage of modifier, and extraction temperature exhibited smaller influences. The best conditions with the described PFE instrumentation were 0.95 g/mL, 10% methanol, and 55°C. An analysis of interactions among variables gave no significant interactions.

Catechin was selected as the quantitative target analyte in this study. It was chosen because of its high sensitivity to degradation processes. We also reasoned that it should be one of the polyphenols with a high variability in extraction recovery. Catechin was identified by retention factor and by the addition of an authentic standard of catechin to the extract. The chromatographic peak area of catechin was measured and divided first by the chromatographic peak area of the internal standard (β-resorcyclic acid) and then the amount of seed sample. Results (i.e. area ratio/gram of seed) for each of the five extractions revealed a relative standard deviation of 7.3%.

It was of interest to compare the efficiency of the PFE method with the conventional extraction method using organic solvent. Several methods for extracting catechins from grape seed are available in the literature (12-16). The most general methodology (i.e. liquid solid extraction, LSE) involved the use of aqueous methanol for 16-24 hours at room temperature. A sonicated assisted liquid solid extraction (SALSE) was also attempted. The time used for SALSE matched the times used for PFE. The results are shown in Table II expressed as milligrams of catechin extract per 100 grams of seed. The LSE and SALSE results are statistically the same while, PFE produced higher recoveries (e.g. 16-20% greater) than LSE or SALSE. The differences in catechin recovery from grape seeds may be due to insufficient solvating power of the aqueous methanol or due to degradation processes during the time of extraction. Reproducibility,

Table II. Comparison of Recoveries for Three Extraction Methods (mg of catechin per 100 g of seeds)[b]

	LSE*	SALSE**	PFE***
mean	65.6[a]	63.0[a]	77.6
RSD	3.3	3.9	7.3
Rel. Rec. (PFE)	84.5	81.2	100.0

[a]Statistically lower than recovery obtained by SFE (t-test, 95% confidence level)

*LSE: Liquid Solent Extraction

**SALSE: Sonicated Assisted Liquid Solvent Extraction

***PFE: Pressurized Fluid Extraction

[b]Reproduced with permission from Journal of Chromatography, 849, (1999) 117-124.

however, with LSE or SALSE was better than PFE. One advantage of PFE over LSE or SALSE is that full automation is feasible by coupling PFE with a chromatographic assay.

Once an optimized extraction procedure had been developed for polyphenols, our attention turned to nonpolar grape extractables. Since 100% CO_2 was to be used, much less experiemntal optimization was required. Larger samples were desired since the isolated fractions would be subjected to off-line spectrometric analysis. Two fractions were created. Eight extractions were performed on 1.5g sample. The average yield was 159.3 mg (10.6%) for the first fraction (fraction A) and 118.4 mg (7.9%) for the second fraction (fraction B). The lowest polarity fraction (A), was a yellow oil after drying under a stream of nitrogen. The Wiley library was used to identify the extracted compounds. There were three different kinds of compounds isolated in the oil fraction, i.e. aliphatic aldehydes, fatty acids and their derivatives, and sterols (Figure 3). The most abundant compounds were the fatty acids, mainly linoleic and oleic acid. Other fatty acids appear in lower amounts, such as palmitic and stearic acids. The first eluants in gas chromatography were the aliphatic aldehydes: 2-*trans*-heptanal, nonanal, 2-*trans*-decenal, 2-*trans*-4-cis-decadienal. These compounds play an important role in the aroma of some oils *(19,20)*. The composition of our fraction A via PFE was, therefore, similar to the composition of grape seed oil obtained by LSE with an organic solvent.

Extraction of the same seeds a second time using modified CO_2 yielded fraction B which matched very well the previous polyphenol extracts (major components = catechin and epicatechin, minor component = gallic acid). These components were identified by their retention times against standard samples. On the basis of the chromatographic conditions employed (i.e. column, elution and detection properties), the other minor compounds are, most probably, phenolic compounds.

Bioassays were performed on the two fractions. As can be seen in Table III, the highest activities exhibited by fraction A were against *S. coagulans niger, E. cloacae, C. freundii* and *E. coli*. These bacteria were susceptible at the concentration assayed. On *S. aureus* there was only moderate activity at the highest concentration. *B. cereus* and the fungus *A. flavus* were resistant to fraction A at all concentrations assayed. Activities on human pathogens by fraction B were lower that those exhibited by fraction A. Only at the highest concentration was there some moderate activity. However, the activity profile was more homogeneous than in the case of fraction A (i.e. the activity was almost the same on all bacteria, with the exception of *B. cereus* which was resistant at all the concentrations assayed). The activity demonstrated by catechin was nil for all bacteria with the exception of *E. coli*, which exhibited moderate activity. For epicatechin, the resulting activity profile was similar to the profile obtained by fraction B, so it is possible that the antibacterial activity of fraction B is mainly due to epicatechin. There were no effects with fractions

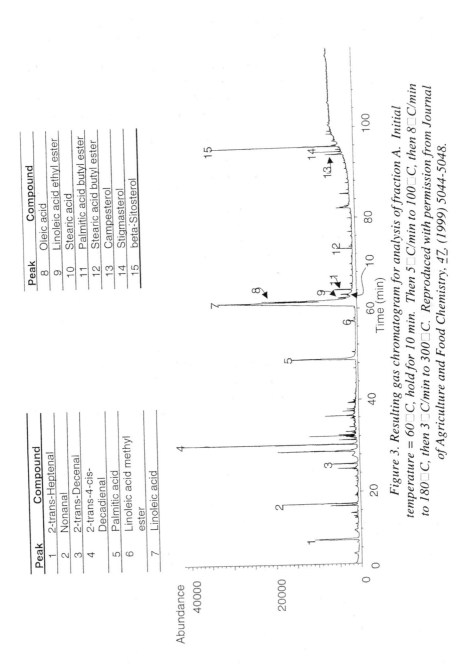

Peak	Compound
1	2-trans-Heptenal
2	Nonanal
3	2-trans-Decenal
4	2-trans-4-cis-Decadienal
5	Palmitic acid
6	Linoleic acid methyl ester
7	Linoleic acid

Peak	Compound
8	Oleic acid
9	Linoleic acid ethyl ester
10	Stearic acid
11	Palmitic acid butyl ester
12	Stearic acid butyl ester
13	Campesterol
14	Stigmasterol
15	beta-Sitosterol

Figure 3. Resulting gas chromatogram for analysis of fraction A. Initial temperature = 60 □C, hold for 10 min. Then 5 □C/min to 100 □C, then 8 □C/min to 180 □C, then 3 □C/min to 300 □C. Reproduced with permission from Journal of Agriculture and Food Chemistry, 47, (1999) 5044-5048.

on fungal phytopathogens (Table III), with the exception of moderate activity of fraction A on *C. echinulatum.*

It can be concluded that by PFE it is possible to fractionate compounds from grape seeds. The results of the bioassays mean that there are active compounds in fractions extracted by PFE. More work should be done in order to determine the compounds responsible for activity in fraction A against bacteria and the responsible compounds of fraction B in the etiolated wheat coleoptile bioassay. The gentle conditions used and the absence of air and light during the PFE process guarantee the conservation of bioactivities and antioxidant power of extracted compounds because degradation processes are avoided.

Table III. Effects of Fraction A and Fraction B Against Human Pathogen Microorganisms[b]

organism	Fraction A			Fraction B		
	μg/disk[a]					
	1500	1000	500	1500	1000	500
Bacillus cereus (+)	-	-	-	-	-	-
Staphylococcus aureus (+)	+	-	-	+	+	-
Staphylococcus coagulans niger (+)	++	++	++	+	+	-
Citrobacter freundii (-)	++	+	-	+	+	-
Echerichia cloacae (-)	++	++	+	+	+	+
Echerichia coli (-)	+	+	-	+	+	-
Aspergillus flavus	-	-	-	-	-	-

[a]4-mm disk impregnated with selected concentration of fraction A or B.

++ high inhibition (diameter > 15 mm)

+ inhibition (15 mm > diameter > 10 mm)

- no inhibition (diameter < 10 mm)

[b]Reproduced with permission from Journal Agriculture and Food Chemistry, 47, (1999) 5044-5048.

Acknowledgements

We are most appreciative to ISCO-Suprex for the loan and maintenance of extraction instrumentation, to Air Products and Chemicals Inc. for the donation of high quality carbon dioxide and to Synthon Inc. for providing the grape seeds. R. M. Varela is acknowledged for providing the bioassays. M. Palma gratefully acknowledges the University of Cádiz for the award of a sabbatical fellowships.

References

1. Revilla, E.; Alonso, E.; Kovak, V. in *Wine: Nutritional and Therapeutic Benefits*. T. R. Watkins (Ed.) ACS Symposium Series *661*, **1997**, 69-80.
2. Smith, A. K.; June, H.; Noble, A. C. *Food Qual. Preference* **1996**, *7*, 161-166.
3. Singleton, V. L. in *Plant Polyphenols: Synthesis, Properties, Significance* R. W. Heminway and P. E. Laks (Eds.) Plenum Press: New York, 1992, 859-880.
4. Thorngate, J. H. in *Beer and Wine Production: Analysis, Characterization and technological Advances* B. H. Gump, Ed. ACD Symposium Series *536*, 1993, 51-63.
5. Meyer, A. S.; Yi, O. S.; Pearson, D. A.; Waterhouse, A. L.; Frankel, E. N. *J. Agric Food Chem.* **1997**, *45*, 1638-1643.
6. Saito, M.; Hosoyama, H.; Ariga, T.; Kataoka, S.; Yamaji, N. *J. Agric. Food Chem.* **1998**, *46*, 1460-1464.
7. Liu, L.; Castonguay, A. *Carcinogenesis* **1991**, *12*, 1203-1308.
8. Liverio, L.; Puglisi, P. P.; Morazzoni, P.; Bombardelli, E. *Fitoterapia* **1994**, *65*, 203-209.
9. Takechi, M.; Tanaka, Y.; Nonaka, G. I.; Nishioka, I. *Phytochemistry*, **1985**, *24*, 2245-2250.
10. Vinson, J. A.; Dabbagh, Y. A.; Serry, M. M.; Jang, J. *J. Agric Food Chem.* **1995**, *43*, 2800-2802.
11. Bachi, D.; Garg, A.; Krohn, R. L.; Bachi, M.; Bachi, D. J.; Balmorri, J.; Stohs, S. J. *Gen. Pharmac.* **1998**, *30*, 771-776.
12. Escibano-Bailón, T.; Gutiérrez-Fernández, Y.; Rivas-Gonzalo, J.; Santos-Buelga, C. *J. Agric. Food Chem.* **1992**, *40*, 1794-1799.
13. Fuleki, T.; da Silva, J.M.R. *J. Agric. Food Chem.* **1997**, *45*, 1156-1160.
14. Cork, S. J.; Krockenberger, A. K. *J. Chem. Ecol.* **1991**, *17*, 123-134.
15. Kallithraka, S.; Bakker, J.; Clifford, M. N. *J. Agric. Food Chem.* **1997**, *45*, 2211-2216.

16. Vernhet, A.; Pellerin, P.; Prieur, C.; Osmianski, J.; Moutounet, M. *Am. J. Enol. Vitic.* **1996**, *47*, 25-30.
17. Tipsrisukond, N.; Fernando, L. N.; Clark, A. D. *J. Agric. Food Chem.* **1998**, *446*, 4329-4333.
18. Molero, A.; Pereyra, C.; de la Ossa, E. M. *Chem. Eng. J.* **1996**, *61*, 227-231.
19. Kiritsakis, A. K. *J. Am. Oil Chem. Soc.* **1998**, *75*, 673-691.
20. Shimoda, M.; Nakada, Y.; Nakashima, M.; Osajima, Y. in *Quantitataive Comparison of Volatile Flavor Compounds in Deep-roasted and Light-roasted Sesame Seed Oil. J. Agric. Food Chem.*, **1997**, *45*, 3193-3196..

Chapter 13

Analysis of Polyphenol Constituents in Cocoa and Chocolate

Midori Natsume[1], Naomi Osakabe[1], Toshio Takizawa[1], Tetsuo Nakamura[1], Haruka Miyatake[2], Tsutomu Hatano[2], and Takashi Yoshida[2]

[1]Functional Foods Research and Development Labs., Meiji Seika Kaisha Ltd., Chiyoda, Sakado, Saitama 350–0289, Japan
[2]Faculty of Pharmaceutical Sciences, Okayama University, Tsushima, Okayama 700–8530, Japan

The antioxidant components of cacao liquor, a major ingredient of chocolate and cocoa, have been identified as flavan-3-ols (catechin and epicatechin) and procyanidin oligomers (procyanidin B2, procyanidin C1, cinnamtannin A2, galactopyranosyl-*ent*-(-)-epicatechin (2α→7, 4α→8)-(-)-epicatechin (Gal-EC-EC)). We characterized the procyanidin profiles in cacao liquor, dark chocolate and pure cocoa powder. Furthermore, the effect of the milk on the determination of polyphenols in milk chocolate and milk cocoa was investigated.

It is well known that free radical induced oxidative stress is an important etiologic factor in many pathological processes. Much attention has been focused on the antioxidants in foods. The seeds of *Theobroma cacao* L. (Sterculiaceae) are well known to be rich in polyphenol (*1*). Recently, some antioxidant components of cacao liquor prepared from fermented and roasted cacao beans, which is a major ingredient of cocoa and chocolate products, have been characterized as flavan-3-ol and proanthocyanidin oligomers (*2,3*). In view of their functional effects mentioned before, it is important to determine the composition and quantity of individual constituents (flavan-3-ols and proanthocyanidins) in various cacao products. However, few systematic studies on these constituents have been reported to date. In the present study, qualitative and quantitative analyses by HPLC and LC-MS were performed to examine the proanthocyanidin oligomers, and the individual polyphenols in several cacao liquor preparations derived from cacao beans imported from six different countries, and products (chocolate and cocoa). Furthermore, we investigated the effect of the milk in such products on the determination of polyphenols, as most of the cacao products available on the market contain milk.

Materials and Methods

Isolation and Preparation of Proanthocyanidins from Cacao Liquor

Preparation of Proanthocyanidins from Cacao Liquor

Cacao liquor (1.0 Kg) prepared from cacao beans (1.9 Kg) from Ecuador was treated with a 5-fold volume of *n*-hexane at room temperature for 30 min to remove fats. The defatted cacao liquor (875 g) was then extracted with 70% acetone (6 L), and the acetone was removed by evaporation under reduced pressure. The resulting aqueous solution was extracted with a 9-fold volume of *n*-BuOH. The *n*-BuOH layer after concentration was applied to a Diaion HP-2MG column (15 cm x 10 cm i.d.) and the column was washed with 15% (v/v) ethanol containing 0.1% trifluoroacetic acid in H_2O to remove theobromine. Elution with 80% (v/v) methanol yielded a proanthocyanidin-rich fraction (17 g), which is herein referred to as "cacao liquor proanthocyanidin (CLPr)".

Approximately 10 mg of CLPr was applied to semi-preparative HPLC (normal-phase) using a Supelcosil LC-Si column (25 cm x 21.1 mm i.d., 5 µm). Elution under the conditions described below yielded four fractions: flavan-3-ols (catechins), proanthocyanidin dimers, proanthocyanidin trimers, and proanthocyanidin tetramers, respectively.

Isolation of Proanthocyanidin from Cacao Liquor

Cacao liquor from Ghana was treated with a five-fold volume of *n*-hexane to removed fats, and then the defatted cacao liquor was extracted with 70% acetone. The acetone was removed by evaporation under reduced pressure. The resulting aqueous solution was first extracted with diethyl ether and then with AcOEt. The AcOEt extract was concentrated. It was suspended with EtOH and centrifuged. The supernatant was applied to a Sephadex LH20 column, eluted with EtOH, and four fractions were obtained. Each fraction (A, B, C and D) was applied to a MCIGel CHP20P column, and eluted with H_2O/MeOH in a gradient mode. Polyphenolic substances isolated from each fraction were as follows, catechin from fraction A, epicatechin from fraction B, procyanidin B2 and galactopyanosyl-*ent*- (-)-epicatechin ($2\alpha{\rightarrow}7$, $4\alpha{\rightarrow}8$)-(-)-epicatechin (Gal-EC-EC) from fraction C and procyanidin C1 from fraction D. The aqueous solution, which had been extracted with AcOEt, was extracted with *n*-BuOH. The *n*-BuOH extract was concentrated under reduced pressure. The extract was then applied to a Diaion HP20 column and eluted with mixtures of H_2O/MeOH stepwise. The fraction eluted with 40% aqueous MeOH was concentrated, applied to a MCI gel CHP20P column, and eluted with H_2O/MeOH gradient. From this 40% (v/v) MeOH fraction, cinnamtannin A2 was isolated. The compounds isolated were identified by direct comparison with authentic specimens and/or comparison with data reported in the literature (*4*). The structures of the polyphenols are shown in Figure 1.

Preparation of Polyphenol Solution (PS) from Cacao Liquor, Pure Cocoa and Dark Chocolate

Meiji Seika Kaisha Ltd. Prepared six kinds of cacao liquor using cacao beans imported from six different countries, Ecuador, Venezuela, Ghana, Colombia, the Ivory Coast and Brazil. Pure cocoa and dark chocolate made from cacao liquor, sugar, cacao butter and lecithin were purchased in market.

Each cacao liquor (10.00 g) described above was triturated 3 times with a 5-fold volume of *n*-hexane at room temperature for 30 min to remove most of the fats. The defatted cacao liquor (0.5000 g) was extracted 3 times with a 100-fold volume of 80% (v/v) acetone at 80°C. This aqueous acetone solution, which contained almost all of the catechins and proanthocyanidins, was diluted to 200 mL with 80% acetone in a measuring flask. This solution [polyphenol solution (PS)] was used for total polyphenol analysis and HPLC analysis. Pure cocoa and dark chocolate were also similarly treated to obtain a polyphenol solution (200 mL) derived from each.

Determination of Total Polyphenol Content

The total polyphenol content of the PSs derived from the cacao liquor and cacao products was determined by the Prussian blue method (5), using epicatechin as the standard. The results are expressed as epicatechin equivalent values.

HPLC Analysis

Semi-preparative HPLC

This was performed using a Supelcosil LC-Si column (25 cm x 21.2 mm i.d., 5 μm) with the solvent system CH_2Cl_2-MeOH-HCOOH-H_2O (A) 5:43:1:1 (v/v) and (B) 41:7:1:1 (v/v). Elution was performed using a linear gradient from 0 to 20% A in 40 min, and from 20 to 100% A in 5 min (flow rate, 20 mL/min).

Analysis of Proanthocyanidins

Proanthocyanidin oligomers were analyzed according to the method by Rigaud et al. (6) with slight modification. The conditions are shown in Figure 2.

Analysis of Catechins by Reversed-Phase HPLC (RP-LC)

The HPLC conditions are shown in Figure 2.

Analysis of Proanthocyanidins by Reversed-phase HPLC-Mass Spectrometry (RP-LC/MS)

The HPLC conditions are shown in Figure 3. The RP-LC.MS conditions were described in detail in a previous report (7).

Procyanidin B2

(-)-Epicatechin

(+)-Catechin

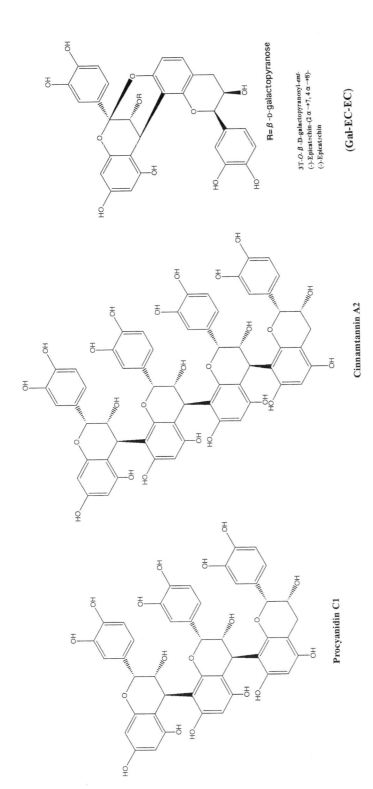

(Gal-EC-EC)

R= β -D-galactopyranose

3T-*O*- β -D-galactopyranosyl-*ent*-
(-)-Epicatechin-(2 α →7, 4 α →8)-
(-)-Epicatechin

Cinnamtannin A2

Procyanidin C1

185

Figure 1. Structures of polyphenols isolated from cacao liquor.

(SOURCE: Reproduced with permission from reference 7. Copyright 2001 American Society for Nutritional Sciences.)

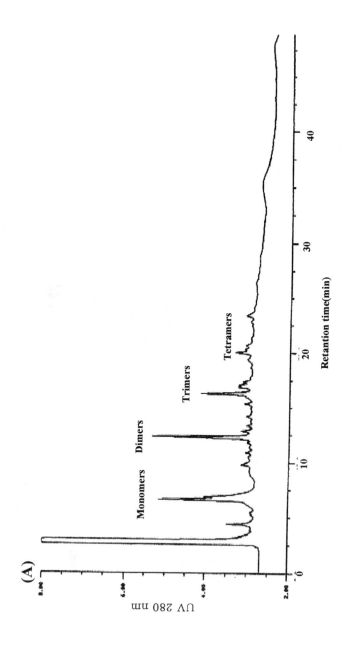

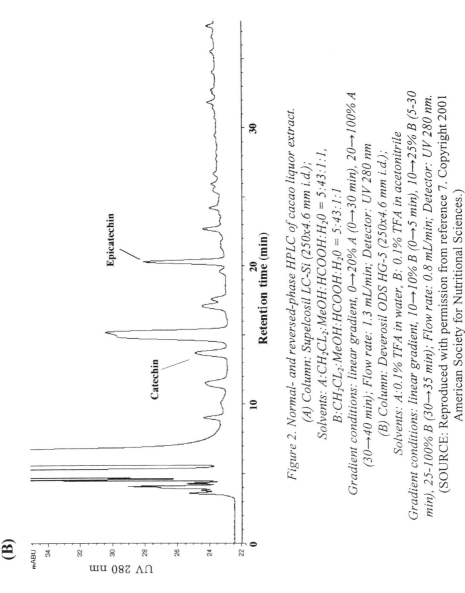

Figure 2. Normal- and reversed-phase HPLC of cacao liquor extract.
(A) Column: Supelcosil LC-Si (250x4.6 mm i.d.);
Solvents: A:CH₂CL₂:MeOH:HCOOH:H₂0 = 5:43:1:1,
B:CH₂CL₂:MeOH:HCOOH:H₂0 = 5:43:1:1
Gradient conditions: linear gradient, 0→20% A (0→30 min), 20→100% A
(30→40 min); Flow rate: 1.3 mL/min; Detector: UV 280 nm
(B) Column: Deverosil ODS HG-5 (250x4.6 mm i.d.);
Solvents: A:0.1% TFA in water, B: 0.1% TFA in acetonitrile
Gradient conditions: linear gradient, 10→10% B (0→5 min), 10→25% B (5-30
min), 25-100% B (30→35 min); Flow rate: 0.8 mL/min; Detector: UV 280 nm.
(SOURCE: Reproduced with permission from reference 7. Copyright 2001
American Society for Nutritional Sciences.)

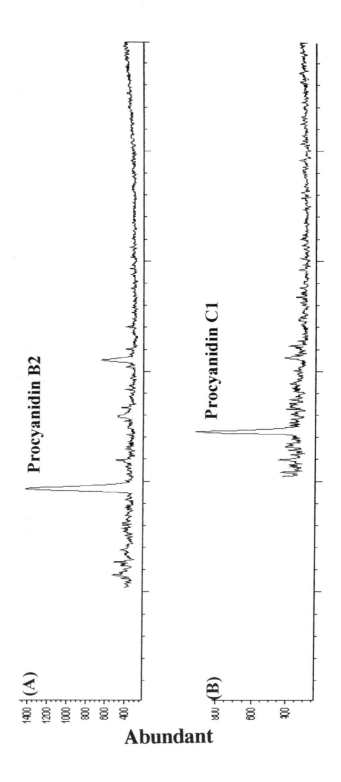

Figure 3. Chromatograms obtained by RP-LC/MS of cacao liquor extract.
(A) (M-H)⁻ =577, (B) (M-H)⁻ =865, (C) (M-H)⁻ =1153, (D) (M-H)⁻ =737; HPLC
conditions: Column: CAPCELL PAK UGI20 (150x2.0 mm i.d.); Solvent:
A:0.03%formic acid in water; B: acetonitrile; Gradient condition:linear
gradient. 10-59%B (0-30 min); Flow rate: 0.2 mL/min; Mass conditions:
Detector: HP1100 mass selective detector; Mode: elecrospary ionization,
selected ion monitoring (SIM), negative.
(SOURCE: Reproduced with permission from reference 7. Copyright 2001
American Society for Nutritional Sciences.)

Continued on next page.

190

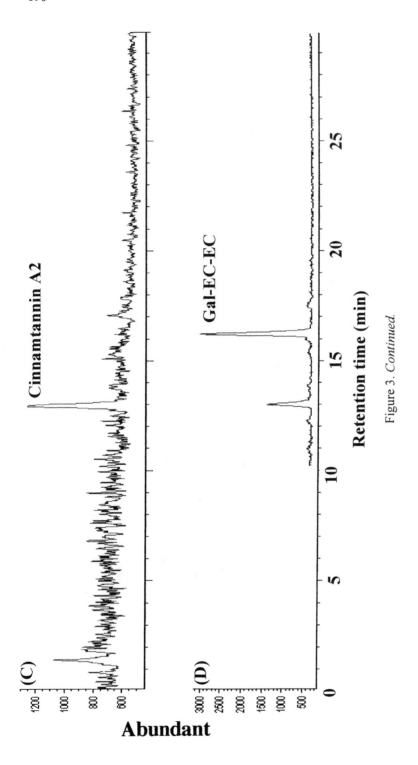

Figure 3. *Continued.*

Effect of Milk Protein on Measurement of the Polyphenol Content of Milk-Containing Cacao Products

Sample

The products examined were prepared in our laboratories. Chocolate samples, except for the case of sample 1, were conching ingredients prepared as follows: sample 1, cacao liquor 100%; sample 2, cacao liquor, 24.39%, cacao butter, 32.93%, sucrose, 42.68%; sample 3, cocao liquor, 20.0%, cacao butter, 27%, sucrose, 35.0%, whole dried milk, 18.0%. Conching is one of the processes involved in making chocolate. It is a process in which cacao liquor and other ingredients are mixed while the acidic flavor is removed.

Cocoa samples, except for the case of sample, were granulated ingredients prepared as follows: sample 4, cocoa powder, 100%; sample 5, cocoa powder, 50.0%, sucrose, 50.0%; sample 6, cocoa powder, 33.3%, sucrose, 33.3%, whole dried milk, 33.3%.

Preparation of a Polyphenol Solution from Each of the Prepared Samples

Polyphenols were extracted from these chocolate and cocoa samples by the above method, and the polyphenol solutions obtained were used for total polyphenol analysis and HPLC analysis.

Results and Discussion

Chromatograms Obtained by NP-LC, RP-LC and RP-LC/MS

Typical chromatograms obtained in analysis of the cacao liquor extract by NP-LC, RP-LC and RP-LC/MS are shown in Figures 2 and 3.

Evaluation of the Recovery of Added Standards

To the defatted cacao liquor (ca. 0.5 g), CLPr as a source of proanthocyanidins was added as a standard in the following amounts: 0, 3.125, 6.25, 12.5, 25.0, 50.0 mg. Each sample was extracted 3 times with a 100-fold volume of

80% (v/v) acetone. The extracted solution was transferred to a 200-mL measuring flask filled with 80% (v/v) acetone, and analyzed by NP-LC and RP-LC. The polyphenol composition of CLPr as standards is shown in Table I. For each proanthocyanidin, the percent recovery was calculated on the basis of duplicate determinations and the results are shown in Table II.

Table I. Profile of CLPr

		Content (%)
Total Polyphenols		72.32
NP-LC	Monomers	10.39
	Dimers	12.07
	Trimers	13.36
	Tetramers	6.24
RP-LC	Catechin	2.49
	Epicatechin	5.89
RP-LC/MS	Procyanidin B2	3.93
	Procyanidin C1	2.58
	Cinnamtannin A2	3.17
	Gal-EC-EC	0.48

*Each value represents the mean of duplicated analyses
SOURCE: Reproduced with permission from reference 7. Copyright 2001 American Society for Nutritional Sciences.

The percent recovery in the case of total polyphenols, epicatechin, procyanidin B2, procyanidin C1, cinnamtannin A2, Gal-EC-EC and procyanidin oligomers was in the range of 83.4 to 119.1%. The percent recovery in the case of catechin was in the range of 68.2 to 98.3%. The recovery of (+)-catechin was low probably due to somewhat poor solubility in the solvent used for extraction in the test examining the recovery of the various compounds.

Reproducibility of Extraction and Analysis

The reproducibility was evaluated by performing six runs of extraction and analysis of the same cacao liquor preparation. The results are shown in Table III.

The reproducibility, except for that in the case of the proanthocyanidin tetramers, was satisfactory. The reproducibility in the case of proanthocyanidin tetramers was poor, probably due to the small content.

Table II. Recovery of Added Standards*

Amount of Added CLPr (mg)	Total Polyphenols (%)	NP-LC			
		Monomers (%)	Dimers (%)	Trimers (%)	Tetramers (%)
100.0	100.0	100.0	100.0	100.0	100.0
3.125	97.8	100.8	99.5	106.6	89.1
6.25	102.2	112.1	99.2	102.9	108.4
12.5	97.4	92.7	105.2	91.8	98.9
25	89.7	90.1	94.1	97.0	113.1
50.0	84.6	83.4	88.6	104.8	104.2

RP-LC			RP-LC/MS		
Catechin (%)	Epicatechin (%)	Procyani-din B2 (%)	Procyani-din C1 (%)	Cinnam-tannin A2 (%)	Gal-EC-EC (%)
100.0	100.0	100.0	100.0	100.0	100.0
98.3	104.6	100.4	99.0	106.4	99.7
95.3	104.4	119.1	106.9	114.5	93.8
68.2	87.6	107.1	103.7	111.6	96.1
80.8	88.6	99.6	98.2	113.0	87.6
92.3	111.5	102.0	87.5	109.7	106.7

*Each value represents the mean of duplicated determinations.
SOURCE: Reproduced with permission from reference 7. Copyright 2001
American Society for Nutritional Sciences.

Table III. Reproducibility of Extraction and Analyses (n=6)

		Value (%)	C.V. (%)
NP-LC	Total Polyphenol	6.63 ± 0.185	2.79
	Monomers	0.454 ± 0.0211	4.65
	Dimers	0.319 ± 0.0108	3.38
	Trimers	0.350 ± 0.0133	3.81
	Tetramers	0.285 ± 0.0190	6.68
RP-LC	Epicatechin	0.0693 ± 0.0023	3.35
	Catechin	0.325 ± 0.0063	1.94
RP-LC/MS	Procyanidin B2	0.170 ± 0.0063	3.70
	Procyanidin C1	0.155 ± 0.0077	4.96
	Cinnamtannin A2	0.281 ± 0.098	3.49
	Gal-EC-EC	0.0260 ± 0.0009	3.41

SOURCE: Reproduced with permission from reference 7. Copyright 2001
American Society for Nutritional Sciences.

Analyses of Polyphenols in Cacao Liquor, Pure Cocoa and Dark Chocolate

The content of polyphenolic substances in cacao liquor preparations derived from cacao beans imported from six different countries is shown in Table IV. According to the NP-LC results, the relative abundance of each type of oligomer except in the case of the cacao beans from Venezuela was as follows: monomers ≥ trimers > dimers ≥ tetramers. The relative abundance of the individual components as determined by RP-LC was as follows: epicatechin > cinnamtannin A2 > procyanidin B2 > procyanidin C1 > catechin > Gal-EC-EC. Comparing these six cacao liquor preparations, the relative proportions of the proanthocyanidin components in each were almost the same, whereas the quantity of each proanthocyanidin component differed markedly. Previous reports have suggested that this type of difference might be due to differences in the kinds of beans used or differences in the fermentation process (8-9).

Table IV. Polyphenol Content of Six Kinds of Cacao Liquor Derived from Cacao Beans Imported from Different Countries*

Cacao Liquor	Total Polyphenols (%)	NP-LC			
		Monomers (%)	Dimers (%)	Trimers (%)	Tetramers (%)
Ecuador	4.11	0.366	0.295	0.344	0.260
Venezuela	1.55	0.106	0.087	0.128	0.101
Ghana	2.93	0.249	0.185	0.215	0.184
Columbia	1.20	0.113	0.076	0.098	0.043
Ivory Coast	3.13	0.230	0.178	0.233	0.241
Brazil	6.04	1.132	0.788	0.787	0.576

RP-LC				RP-LC/MS	
Catechin (%)	Epicatechin (%)	Procyanidin B2 (%)	Procyanidin C1 (%)	Cinnamtannin A2 (%)	Gal-EC-EC (%)
0.040	0.227	0.124	0.080	0.144	0.018
0.014	0.074	0.034	0.026	0.064	0.006
0.027	0.137	0.058	0.041	0.089	0.013
0.020	0.059	0.024	0.012	0.034	0.004
0.017	0.125	0.061	0.053	0.138	0.016
0.063	0.577	0.197	0.151	0.315	0.038

*Each value represents the mean of duplicated determinations.
SOURCE: Reproduced with permission from reference 7. Copyright 2001 American Society for Nutritional Sciences.

The results of analyses of six dark chocolates obtained from the market are shown in Table V. The proportion of proanthocyanidin oligomers was found to be similar to that in the case of cacao liquor. The results of analyses of six pure cocoas obtained from the market are shown in Table VI. The relative abundance of proanthocyanidins in the cocoas was monomers > dimers > trimers > tetramers as determined by NP-LC, and the relative abundance of individual components was epicatechin ≥ catechin > procyanidin B2 > cinnamtannin A2 > procyanidin C1 > Gal-EC-EC as determined by RP-LC.

Table V. Polyphenol Content of Six Kinds of Dark Chocolates*

Dark Chocolate No.	Total Polyphenols (%)	NP-LC			
		Monomers (%)	Dimers (%)	Trimers (%)	Tetramers (%)
1	1.59	0.168	0.079	0.159	0.073
2	1.52	0.106	0.076	0.159	0.066
3	2.13	0.200	0.114	0.190	0.083
4	1.23	0.099	0.059	0.133	0.058
5	2.11	0.193	0.142	0.185	0.112
6	1.71	0.142	0.073	0.154	0.059

RP-LC				RP-LC/MS	
Catechin (%)	Epicatechin (%)	Procyanidin B2 (%)	Procyanidin C1 (%)	Cinnamtannin A2 (%)	Gal-EC-EC (%)
0.027	0.067	0.034	0.022	0.040	0.004
0.022	0.072	0.032	0.024	0.056	0.005
0.038	0.122	0.046	0.033	0.065	0.006
0.023	0.050	0.021	0.013	0.029	0.004
0.032	0.124	0.054	0.044	0.086	0.008
0.026	0.085	0.032	0.020	0.047	0.004

*Each value represents the mean of duplicated determinations.

The total polyphenol content of the cocoas was higher than that of the chocolates, while the amounts of proanthocyanidin oligomers, except for the monomer, in the cocoas were lower than those in the chocolates. Also, pure cocoa powder differed from cacao liquor and chocolate in terms of the following ratios: proanthocyanidin dimer/trimer, catechin/epicatechin, and procyanidin B2/cinnamtannin A2. Chocolate is produced by mixing cacao liquor and sugar, and molding. On the other hand, cocoa is produced through complicated processes including treatment with alkaline, removal of lipids by means of a press at high temperature, and milling to specified particle sizes. The differences in the proanthocyanidin profile and quantity between chocolate and cocoa might thus be explained by the difference in manufacturing methods, as alkalization is known to lead to chemical alterations of polyphenols (2,10).

Table VI. Polyphone Content of Six Kinds of Pure Cocoa*

| Pure Cocoa No. | Total Polyphenols (%) | NP-LC | | | |
		Monomers (%)	Dimers (%)	Trimers (%)	Tetramers (%)
1	3.02	0.315	0.070	0.069	n.d
2	3.50	0.374	0.178	0.114	0.120
3	3.45	0.262	0.118	0.139	0.099
4	3.42	0.290	0.108	0.114	0.097
5	4.54	0.417	0.147	0.105	0.065
6	4.73	0.393	0.138	0.183	0.080

| RP-LC | | | | RP-LC/MS | |
Catechin (%)	Epicatechin (%)	Procyanidin B2 (%)	Procyanidin C1 (%)	Cinnamtannin A2 (%)	Gal-EC-EC (%)
0.092	0.063	0.013	0.005	n.d.	0.002
0.089	0.109	0.051	0.030	0.042	0.006
0.070	0.088	0.037	0.024	0.034	0.005
0.094	0.093	0.036	0.021	0.031	0.005
0.137	0.123	0.045	0.027	0.036	0.005
0.078	0.132	0.057	0.036	0.056	0.007

*Each value represents the mean of duplicated determinations.
SOURCE: Reproduced with permission from reference 7. Copyright 2001 American Society for Nutritional Sciences.

Effect of Milk Protein on Determination of the Polyphenol Content of Cacao Products

The results of evaluation of the recovery of polyphenols in the presence of added sugar and/or milk are shown in Table VII. In this examination, the percent recovery of polyphenols was in the range of 92.3-118.8%. The percent recovery of total polyphenols from chocolate samples was in the range of 125.4-117.5% and that of catechin was in the range of 76.7-84.9%. Milk protein is known to form a complex with polyphenols (11-12). However, in this study, milk protein did not interfere with determination of the polyphenol content of the cacao products containing milk by this method, as 80% acetone was used as the extraction solvent.

Table VII. Recovery of Polyphenols*

Sample		Total Polyphenol (%)	RP-LC	
			Catechin (%)	Epicatechin (%)
Chocolate	Sample 1	100.0	100.0	100.0
	Sample 2	125.4	84.9	118.8
	Sample 3	117.5	76.7	107.5
Cocoa	Sample 1	100.0	100.0	100.0
	Sample 2	107.8	95.4	101.8
	Sample 3	97.5	92.3	103.0

Sample		RP-LC/MS		
		Procyanidin B2 (%)	Procyanidin C1 (%)	Cinnamtannin A2 (%)
Chocolate	Sample 1	100.0	100.0	100.0
	Sample 2	117.0	117.0	112.8
	Sample 3	109.2	108.8	113.1
Cocoa	Sample 1	100.0	100.0	100.0
	Sample 2	120.8	99.3	105.4
	Sample 3	104.4	92.3	94.8

*Each value represents the mean of duplicated determinations.

Conclusions

1. We have established suitable methods of analysis of polyphenolic substances in cacao products.
2. The profile of catechins and proanthocyanidins in cacao products was analyzed with high reproducibility by the methods used.
3. Milk protein did not interfere with determination of the polyphenol content of milk chocolate or milk cocoa by the method used.

References

1. Forsyth, W.G.C. *Biochem.* **1955**, *60*, 108-111.
2. Adamson, G.E.; Lazarus, S.A.; Mitchell, A.E.; Prior, R.L.; Cao, G.; Jacobs, P.H.; Kremers, B.G.; Hammerstone, J.F.; Rucker, R.B.; Ritter, K.A.; Schimitz, H.H. *J. Agric. Food Chem.* **1999**, *47*, 4184-4188.
3. Sanbongi, C.; Osakabe, N.; Natsume, M.; Takizawa, T.; Gomi, S.; Osawa, T. *J. Agric. Food Chem.* **1998**, *46*, 454-457.
4. Porter, L.J.; Ma, Z.; Chan, B.G. *Phytochemistry* **1991**, *30*, 1657-1663.
5. Hagerman, A.E.; Butler, L.G. In *Oxygen Radicals in Biological Systems;* Packer, L. Ed.; *Methods in Enzymology* **1994**, *234*, 432-433.
6. Rigaud, J.; Escribano-Bailon, M.T.; Prieur, C.; Souquet, J.M.; Cheynier, V. *J. Chromatogr.A* **1993**, *654*, 255-260.
7. Natsume, M.; Osakabe, N.; Yamagishi, M.; Takizawa, T.; Nakamura, T.; Miyatake, H.; Hatano, T.; Yoshida, T. *Biosci. Biotechnol. Biochem.* In press.
8. Fowler, M.S. In *Industrial Chocolate Manufacture and* Use; Beckett, S.T. Ed.; Blackwell Science Ltd.: UK, 1999; pp. 8-35.
9. Meursing, E.H.; Zijderveld, J.A. In *industrial Chocolate Manufacture and* Use; Beckett, S.T. Ed.; Blackwell Science Ltd.: UK, 1999; pp. 101-114.
10. Schenkel, H.J. *Manufact. Confect.* **1973**, 53, 26-33.
11. Ricardo-da-Saliva, J.M.; Cheynier, V.; Sonquet, J.-M.; Moutounet, M.; Cabanis, J.-C.; Bourzeix, M. *J. Sci. Food Agric.* **1991**, *57*, 111-125.
12. Haslam, E. *Biochem.J* **1974**, *139*, 285-288.

Chapter 14

A Quantitative HPLC Method for the Quality Assurance of Goldenseal Products in the U.S. Market

Mingfu Wang[1], Nanqun Zhu[2], Yi Jin[2], Nathan Belkowitz[1], and Chi-Tang Ho[2]

[1]Research and Development Department, Quality Botanical Ingredients Inc., 500 Metuchen Road, South Plainfield, NJ 07080
[2]Department of Food Science, Rutgers University, 65 Dudley Road, New Brunswick, NJ 08901–8520

A rapid and quantitative method of quality assurance for the marker phytochemicals and bioactive components in products containing materials derived from Goldenseal has been developed. By using a Luna phenyl-hexyl column with a mixture of sodium lauryl sulfate-phosphoric acid-water-isopropanol-acetonitrile as mobile phase, the two marker compounds, hydrastine and berberine were well separated at ambient temperature in 20 minutes. The alkaloid contents of six commercial available Goldenseal root and Goldenseal/ Echinacea supplements were also analyzed.

Goldenseal (*Hydrastis canadensis* L.) is an herbaceous perennial with a short horizontal rhizome bearing multiple roots, two or three palmatilobate leaves and a unique terminal, greenish-white flower. Over the years, it has also been referred to by a large number of other names, including yellow root, ground raspberry, eye-balm, yellow paint, wild turmeric and yellow eye. The plant grows wild in the eastern part of North America, but it is mostly cultivated now. Goldenseal was used extensively by Native Americans (the Cherokee people) as an herbal medication and clothing dye. Its root has been listed in the US Pharmacopoeia from 1803 to 1926. It is also listed in the 10th edition of the French pharmacopoeia which specify that the part to be used "consists of the dried rhizome and roots and that they must contain not less than 2.5% hydrastine calculated relative to the dried drug (1). However the aerial part is also sold in the USA nutraceutical market in different formula.

Goldenseal is now a top selling herb in North America. Its medicinal uses centered on its ability to soothe the mucous membranes of the respiratory, digestive, and genitourinary tracts in inflammatory conditions induced by allergy or infection. It can be found in many formulations, including treatment of nasal congestion, mouth sores, eye and ear infections, and as a topical antiseptic. It is commonly believed to help boost the immune system, increase the efficacy of other medicinal herbs, and for this purpose, it is usually formulated together with Echinacea.

The isoquinoline alkaloids are the major components in goldenseal root, including hydrastine (1.5-4%) berberine (0.5-6%), canadine, candaline and other minor alkaloids (2-4). Goldenseal also contains fatty acids, resin, polyphenolic acids, chlorogenic acid, meconin and a small amount of volatile oil.

Modern pharmacological research has found that the alcoholic extract of Hydrastis canadensis L. root is well known for its vasoconstrictive effects (1,5). And a number of immunomodulatory effects have been attributed to goldenseal (6) and the alcoholic extract of goldenseal root also showed reversible smooth muscle relaxant activity (7,8). One of the major alkaloids, berberine has a long history of medicinal use in both Ayuvedic and Chinese medicine. It has demonstrated significant anti-microbial activity against various organisms including bacteria, viruses, fungi, protozoans, helminths and chlamydia (1,9). Currently the clinical uses of berberine include bacterial diarrhea, intestinal parasite, and ocular trachoma infection. In addition to similar action, hydrastine is vasoconstrictive, stimulates bile secretion and involuntary muscles in human and display sedative and hypotensive effects (5). Due to the activity of berberine and hydrastine, the quantities of berberine and hydrastine are therefore used as a quality control measure in standardized herbal and nutraceutical preparation.

Although the analysis of berberine in other botanicals have been studies extensively (10-14). Surprisingly, very few groups have ever reported on the quantity of berberine and hydrastine found in herbal raw materials or commercially available dietary supplements containing goldenseal. Leone et al. (15) recently reported an HPLC determination of the major alkaloids extracted from Hydrastis canadensis L, but this method is not validated, the peak separation and shape are not good. Caslavska and Thormann reported in 1998 the analysis of isoquinoline alkaloids in goldenseal by capillary electrophoresis coupled with mass spectrometry (*16*). In this research a new, simple and validated ion-pair high-performance chromatography method for simultaneous determination of berberine and hydrastine (structures shown in Scheme 1) in goldenseal herbs, goldenseal roots and dietary supplements containing goldenseal root extract was developed.

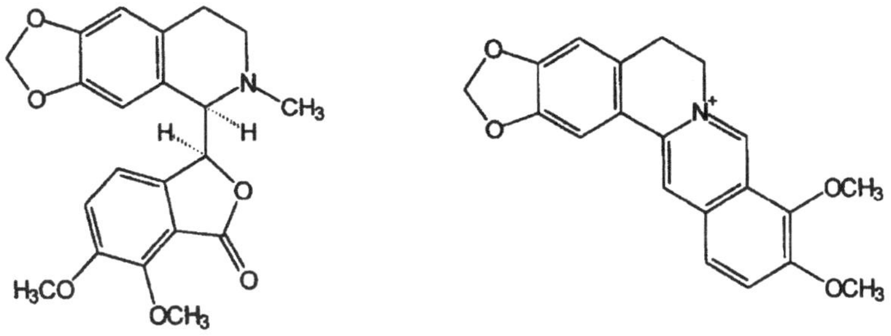

Hydrastine   Berberine

Scheme 1. Structures of hydrastine and berberine

Materials and Methods

Materials

Hyrastine hydrochloride and berberine hydrochloride and sodium lauryl sulfate were purchased from Sigma Chemical Co. (St. Louis, MO). Phosphoric acid 85%, HPLC grade acetonitrile, water and isopropanol were obtained from Aldrich Chemical Co. (Milwaukee, WI). Goldenseal roots and goldenseal herbs (aerial part) were obtained from commercial suppliers in the North American market. Several goldenseal roots, goldenseal/Echinacea dietary supplements were purchased from local retailers in New Jersey.

Instrumentation and Chromatographic Conditions

Analytical HPLC analyses were performed on a Hewlett-Packard 1100 modular system equipped with an autosampler, a quaternary pump system, a photodiode array detector and a HP Chemstation data system. A prepacked 150×4.6 mm (3 μ particle size, 00F-4256-E0) Luna phenyl-hexyl analytical column (Phenomenex, Torrance, CA) were operated with a mobile phase consisting of acetonitrile, isopropanol, water and phosphoric acid (34:10:56:0.2, v/v/v/v). Sodium lauryl sulfate, an ion-pairing agent, was added to the mobile phase to achieve a final concentration of 5 mM. The mobile phase was delivered at a flow rate of 1.2 mL/minute at room temperature. The absorption spectra were recorded from 200 to 400 nm for all peaks; quantitation was carried out at a single wavelength of 235 nm.

Standard Preparation

Mix 250 mL methanol, 250 mL water and 10 mL concentrated HCl to make a 50% methanol acidic extraction solvent. Accurately weigh about 15 mg hydrastine hydrochloride and 15 mg berberine hydrochloride into 25 mL volumetric flask, record the weight (correcting for water and HCl). Add approximately 15 mL 50% methanol acidic solution and sonicate for 15 minutes. Allow the flask to cool to room temperature and fill to full volume with 50% methanol acidic solution. Measure 5 mL above solution and transfer to a 25 mL volumetric flask and diluted to the full volume using 50% methanol acidic solution (standard solution).

Preparation of Samples

For herbs, accurately weigh about 400 mg goldenseal (aerial part), for goldenseal root, accurately weigh 100 mg, for herbal extract (10% total alkaloids), accurately weigh about 50 mg into 50 mL volumetric flask. For goldenseal dietary supplements, weigh 20 capsules or tablets and record the average value for each capsule or tablet, then those capsules and tablets were pulverized and 100 mg of these samples were added into a 50 mL volumetric flask. Add appropriately 35 mL of 50% methanol acidic solution and sonicate for 25 min and shake for 15 min. Allow the flask to cool to room temperature and fill to volume with 50% methanol acidic solution. Using a disposable syringe and 0.45 μm filter, filter the sample into an HPLC vial.

Results and Discussion

A Luna phenyl-hexyl column (150×4.6mm, 3 µ particle size) was selected for the ion-pairing HPLC. Various mobile phase systems were evaluated to achieve a satifactory seperation of berberine and hydratine. A mixture of sodium lauryl sulfate, acetonitrile, isopropanol, water and phosphoric acid (85%) (0.5:34:10:56:0.2, M/v/v/v/v) was found to be the best. Chromatograms for the goldenseal herb, goldenseal root and goldenseal with Echinacea were showed in Figures 1, 2 and 3, respectively. No interfering peaks were noted for *Hydrastis canadensis* L and good resolution was achieved among all the alkaloids. The retention time for hydrastine and berberine were approximately 8.7 and 16.7 minutes, respectively.

System Suitability

System suitability was checked by performing 10 replicate analyses of one goldenseal sample within the same working day and then check the percent relative standard deviation (%RSD) of the retention time and peak area for both peaks. The %RSD of the retention time for hydrastine and berberine were 0.67% and 0.88%, respectively. The % RSD for the peak area of hydrastine and berberine were 0.32% and 0.30%. It indicated this method is suitable for the analysis of goldenseal.

Precision

The precision of the extraction procedure was validated using one goldenseal herb sample. Six samples, weighing about 400 mg, were extracted as described above. An aliquot of each sample was then injected and quantitated. The average amount of hydrastine was 0.286% with an RSD of 2.10%. The average amount of berberine was 0.366% with an RSD of 1.48%. Suggesting the method has excellent precision.

Percent Recovery

The recovery was determined by removing 25 mL of extraction solvent from all 6 extractions done in precision test and replacing it with 25 mL of fresh extraction solvent. These samples were then re-extracted and reanalyzed for hydrastine and berberine. A value of exactly one half (1/2) the value of the original extraction validates 100% recovery.The average amount of hydrastine recovered from the 6 extracts was 0.286% with an RSD of 2.10%. The average amount of hydrastine recovered from the re-extraction of these samples × 2,

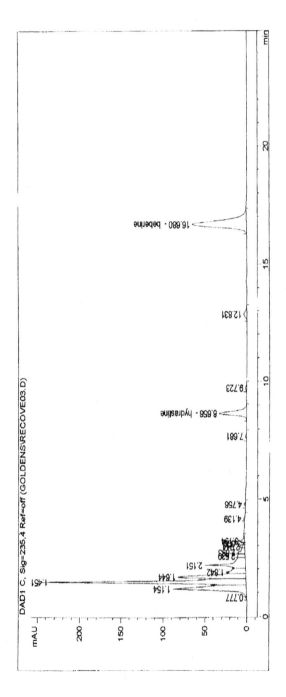

Figure 1. Representative HPLC chromatogram of goldenseal herb (aerial part). Peak at 0.656 min: hydrastine; at 16.680 min: berberine.

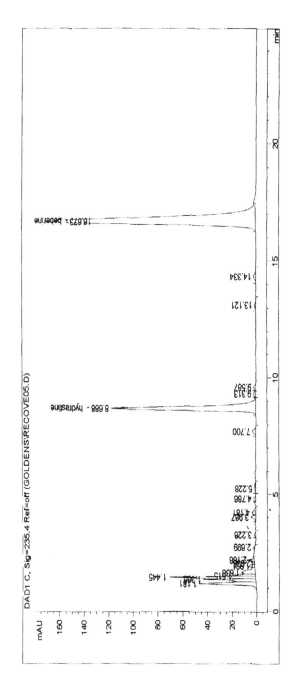

Figure 2. Representative HPLC chromatogram of goldenseal root. Peak at 0.688 min: hydrastine; at 16.673 min: berberine

206

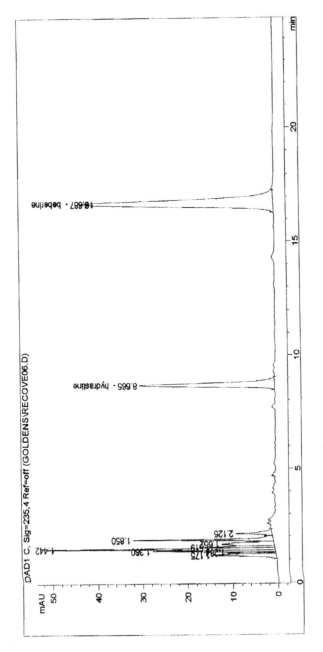

Figure 3. Representative HPLC chromatogram of Goldenseal/Echinacea. Peak at 0.665 min: hydrastine; at 16.687 min: berberine

was 0.278% with an RSD of 0.79%. Assuming the average amount extracted in the re-extraction is 100%, the average amount recovered in the original extraction is then validated to be 102.9%. The average amount of berberine recovered from the 6 extracts was 0.366% with an RSD of 1.48%. The average amount of berberine recovered from the re-extraction of the samples × 2, was 0.372% with an RSD of 1.02%. Assuming the average amount extracted in the re-extraction is 100%, the average amount recovered in the original extraction is then validated to be 98.4%. This proves the excellence of recovery for this method.

Linear Dynamic Range

The linear dynamic range for hydrastine was determined by preparing a standard curve consisting of 9 standard solutions covering a range from 0.001-1.0 mg/mL. The linear dynamic range for berberine was determined by preparing a standard curve consisting of 9 standard solutions covering a range from 0.0006–0.8 mg/mL. The linearity range for hydrastine was determined as 0.002-1 mg/mL with correlation coefficient 0.9996. The linearity range for berberine was determined as 0.002-0.8 mg/mL with correlation coefficient 0.9992 (the calibration curves for hydrastine and berberine are showed in Figures 4 and 5, respectively.). Therefore for hydrastine a result greater than 1.0 mg/mL is considered to be outside the linear range of the assay. The sample must be diluted and reassayed. A result less than 0.002 mg/mL must be reported as negative. For berberine, a result greater than 0.8 mg/mL is considered to be outside the linear range of the assay. The sample must be diluted and reassayed. A result less than 0.002 mg/mL must be reported as negative.

Ruggedness

The ruggedness of the assay was demonstrated by comparing the results obtained from two different chemists. All extractions and standard preparations were done according to the method described above. Two chemists got similar results for one goldenseal sample (one is 5.15% total alkaloids, another is 5.10% total alkaloids).

The Content of Alkaloids in Goldenseal Herbs and Goldenseal Roots

Over twenty goldenseal root and herb samples were tested. Hydrastine and berberine were found to exist in both herbs and roots. But the content were quite different, in herb (aerial part) samples, hydrastine ranged in content from 0.27-0.29%, berberine ranged in content from 0.36-0.39%. While in goldenseal

208

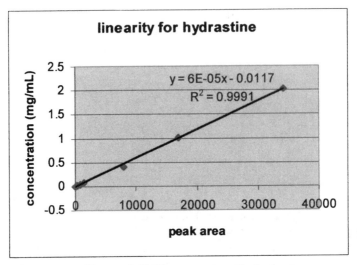

Figure 4. Calibration curve for hydratine

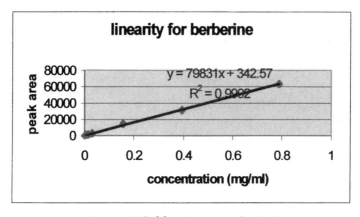

Figure 5. Calibration curve for berberine.

root, hydrastine ranged in the content from 2.25-3.32% and berberine ranged from 2.61% to 3.75%. In the market, goldenseal is also sold as 5% or 10% alkaloid extracts. Most of the extracts we tested met the claim, but half of the extract contained only berberine, hydrastine were either not observed or the content was very low. They are suspected not to be the right species, because goldenseal usually contains almost equal amounts of berberine and hydrastine. On the other hand, berberine exists in quite a number of herbs such as *Coptis chinese, Berberis aquifolium, Berberis vulgaris, Berberiis aristata,* etc. and these herbs usually contain either no or very low content of hydrastine. Representative chromatograms of the wrong species were showed in Figures 6, 7 and 8.

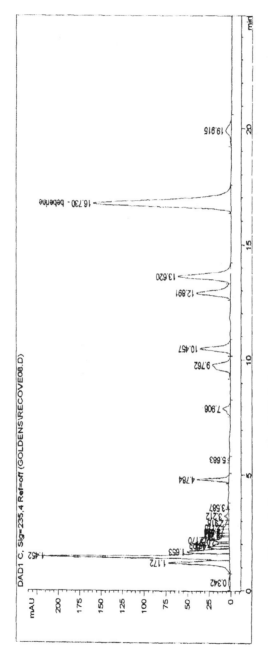

Figure 6. Representative HPLC chromatogram of a wrong species sample A.

210

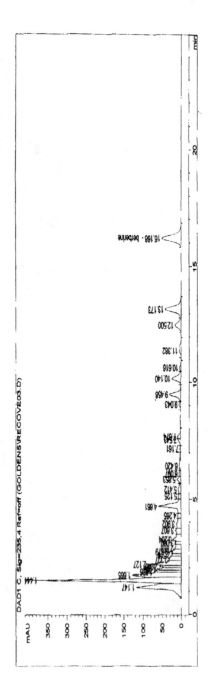

Figure 7. Representative HPLC chromatogram of a wrong species sample B.

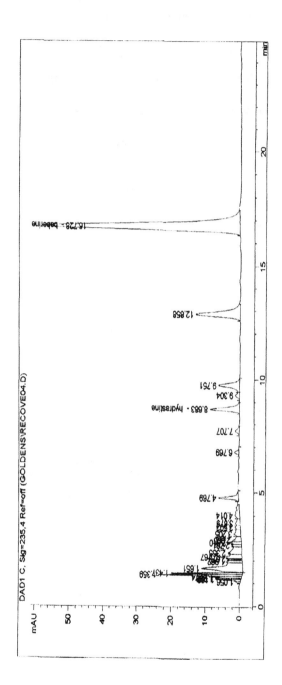

Figure 8. Representative HPLC chromatogram of a wrong species sample C.

212

Evaluation of Commercial Sample (Finished Products)

Although goldenseal has been considered as safe alternatives to conventional drugs for many years, its dosage has not been set. So it is not surprised to notice that the contents of alkaloids are different from products to prodcuts covering a range of 1.25 mg to 57 mg total alkaloid per serve. In the US market, goldenseal is usually formulated together with Echinacea for immune system boosting. We tested the alkaloids contents of three commerical Goldenseal/Echinacea products, no chromatographic interference was noted for Echinacea (*Echinacea angustifolia and Echinacea purperea*) or other ingredients. Two of them matched the label claim for total alkaloids (but one products only contained berberine and hydrastine was not detected), one failed with low alkaloid content. Two goldenseal root products were also tested for the total alkaloids, one was found to contain 55.5 mg alkaloids (calculated as hydrastine and berberine, label claim is 57 mg per serve), the other was found to contain 27.1 mg alkaloids (calculated as hydrastine and berberine, while the label only claimed 10 mg hydrastine/per serve).

Conclusion

A rapid and quantitative method of quality assurance of hydrastine and berberine contents in products containing materials derived from Goldenseal has been developed. By using a Luna phenyl-hexyl column with a mixture of sodium lauryl sulfate-phosphoric acid-water-isopropanol-acetonitrile as mobile phase, this method showed excellent linearity, accuracy and precision. The alkaloid contents of goldenseal roots, goldenseal root extracts, commercial available goldenseal roots and goldenseal/echinacea supplements were also analyzed. For the goldenseal root extracts (5% and 10% alkaloids), quite a lot of samples only contained berberine or the concentrations of hydrastine were very low. These samples may not be the goldenseal.

References

1. Bruneton, J. *Technique & Documentation-Lavoisier* **1995**, 740-741.
2. Gentry, E.; Jampani, H.B.; Keshavarz-Shokri, A; Morton, M.D.; Velde, D.; Telilepalli, H.; Mitscher, L.A.; Shawar, R.; Humble, D.; Baker, W. *Journal of Natural Products* **1998**, *61*, 1187-1193.
3. Gleye, J.; Stanishlas, E. *Plantes Medicinales et Phytotherapie* **1972**, *6*, 306-310.
4. Gleye, J.; Ahond, A.; Stanislas, E. *Phytochemistry* **1974**, *13*, 675-676.

5. Shideman, F.E. *Bull. National Formulary Comm.* **1950**, *18*, 3-9.
6. Rehman, J.; Dillow, J.M.; Carter S.M.; Chou, J.; Le, B.; Maisel A.S. *Echinacea angustifolia* and *Hydrastis canadensis, Immunology Letters,* **1999**, *68*, 391-395.
7. Cometa, M.F.; Galeffi, C.; Palmery, M. *Phytotherapy Research,* **1996**, *10:* Supplement 1, S56-S58.
8. Cometa, M.F.; Abdel-Haq, H.; Palmery, M. *Phytotherapy Research* **1998**, 12: Supplment 1, S83-S85.
9. Anonymous, *Alternative Medicine Review* **2000**, *5*, 175-177.
10. Misaki, T.; Sagara, K.; Ojima, M.; Kakizawa, S.; Oshima, T.; Yoshizawa, H. *Chem. Pharm. Bull.* **1982**, *30*, 354-357.
11. Lin, S.-J.; Tseng, H.-H.; Wen, K .-C.; Suen, T.-T. *J. Chromatography* **1990**, *730*, 17-23.
12. Liu, Y.-M.; Sheu, S.-J. *J. Chromatography* **1992**, *623*, 196-199.
13. Luo, G.; Wang, Y.; Zhou, G.; Yu, Y. *Journal of Liquid Chromatography* **1991**, *13*, 3825-3832.
14. Lee, H.S.; Eom, Y.E.; Eom, D.O. *Journal of Pharmaceutical & Biomedical Analysis* **1999**, *21*, 59-63.
15. Leone, M.G.; Cometa, M.F.; Palmery, M.; Saso, L. *Phytotherapy Research* **1996**, *10*, S45-S46.
16. Sturm, S.; Stuppner, H. *Electrophoresis* **1998**, *19*, 3026-3032.

Chapter 15

Analytical Methods for the Active Components in Tea Products

Huizhen Zhang, Long-Ze Lin, Xian-Guo He, and
Michael P. Petteruti

Research Laboratory of Natural Products, A. M. Todd Company,
150 Domorah Drive, Montgomeryville, PA 18936

The infusion of dried tea leaves is one of the most popular
beverages in the world. Most recently, the relationship
between tea consumption and prevention of certain forms of
human cancer has received a great deal of attention. In order
to control the quality and ensure the reliability and
repeatability of pharmacological and clinical research, it is
necessary to develop validated analytical methods for the
qualitative and quantitative analysis of the active components
found in tea products. Tea leaves are rich in polyphenols,
caffeine, vitamins, and amino acids. These active components
in tea products have various physiological and
pharmacological functions, such as antioxidative,
chemopreventative activity, suppressive effect of uremic toxin
formation, prevention of dental caries, etc. This review will
mainly focus on the analytical methods used to evaluate the
active components in tea products. HPLC, HPLC-MS,
spectrophotometry, near-infrared and other methodologies
will be reviewed.

Tea is consumed worldwide and ranks only second to water as a beverage.
It is prepared from dried leaves of *Camellia sinensis (1)*. Over 300 different

kinds of tea are now produced, but there are only three general forms of tea: the unfermented green tea, the partially fermented oolong tea and the fermented black tea. Green tea is manufactured from fresh leaf. The polyphenol components during black tea manufacturing have been extensively oxidized into theaflavins, thearubigins, and other oligomer (2). The chemical composition of tea leaves varies greatly depending on their origin, age and the type of processing (3). Table I shows the composition of fresh green tea leaf based on the dry weight (2). Polyphenol compounds are the most significant group of tea components, especially certain catechins, such as (-)-epicatechin gallate (ECG), (-)-gallocatechin gallate (GCG) and (-)-epigallocatechin gallate (EGCG). Recent interest in green and black tea polyphenol compounds has increased due to their antioxidant activities and their possible role in the prevention of cancer and cardiovascular diseases (3-13). In order to find a constituent or a group of tea components that are a measure of tea quality, to optimize the degree of fermentation that offers the best quality of tea, and to correlate the health effects of tea with certain tea components, it is necessary to develop validated analytical methods for the qualitative and quantitative analysis of the active components found in tea products. This review will mainly focus on the analytical methods used to evaluate the active components in tea products, such as high-performance liquid chromatograph (HPLC), gas chromatograph (GC), mass spectrometry (MS), spectrophotometry (UV/VIS), near-infrared spectroscopy (NIRS) and capillary electrophoresis (CE).

Table I. Composition of Fresh Green Tea Leaf in % Dry Weight

Polyphenols	36	Carbohydrates	25
Methyl xanthines	3.5	Protein	15
Amino acids	4	Lignin	6.5
Organic acids	1.5	Lipids	2
Carotenoids	<0.1	Chlorophyll, etc.	0.5
Volatiles	<0.1	Ash	5

SOURCE: Reproduced with permission from reference 2. Copyright 1992 Academic Press.

Methods for the Analysis of Total Polyphenol Compounds

Polyphenol compounds is the most interesting group of tea leaf components. From Table I, we can see that about 36% (dry weight) polyphenol compounds in fresh tea leaf, such as catechins, flavonol glycosides and proanthocyanidins (2,14,15). These compounds play important role in terms of health benefits of tea consumption (2).

Spectrophotometric Method

Spectrophotometric method is the most suitable method for determination of total polyphenol compounds due to its nonspecific property. A number of spectrophotometric methods was developed for the determination of total polyphenol compounds in plants and plant extracts, including tea leaves and tea products, such as Folin-Denis method, vanillin method, diazotized amines method and Ferrous-Tartrate method *(16-19)*. Folin-Denis method and Ferrous-Tartrate method are two widely used methods now for evaluation tea leaves and tea extracts. Folin-Denis reagents are mixtures of sodium tungstate, phosphomolybdic acid and phosphoric acid which can form color complex with polyphenol compounds. Ferrous-Tartrate solution is prepared by dissolving ferrous sulfate heptahydrate with sodium potassium tartrate into water. It also can form color complex with polyphenol compounds at pH 7.5 buffer solution.

High Performance Liquid Chromatographic (HPLC) Method

HPLC method can be used for reporting total polyphenol compounds by adding each individual identified compound together. Therefore, the meaning of the result of total polyphenol compounds tested by HPLC is different from the result tested by spectrophotometric method. It is very important to compare the data generated by the same method. There are a lot of HPLC method developed for identification, quantification and purification of polyphenol compounds, especially for catechins and flavonol glycosides. The detailed review will be given in the following sections.

Near-Infrared Spectroscopic (NIRS) Method

Near-infrared spectroscopy analysis is an instrumental method for rapidly and reproducibly measuring physical and chemical properties of samples with little or no sample preparations. Since the NIR spectroscopic analysis offers four principal advantages: speed, simplicity of sample preparation, multiplicity of analyses from a single spectrum, and the intrinsic nonconsumption of the sample *(20)*, it has been widely used to measure constituents of many agricultural commodities and food products as well as on-line analysis at food processing sites *(21)*. Schulz *et al.* (*22*) developed a near-infrared reflectance spectroscopic method for the prediction of total polyphenol compounds and alkaloid compounds in the leaves of green tea. Reference methods used for the calibration of the instrument were HPLC and spectrophotometric method. The study demonstrated that NIRS technology could be successfully applied as a rapid method to estimate quality of green tea and to control industrial processes.

Methods for the Analysis of Catechins

Catechins are a major group of polyphenolic compounds found in the tea leaves, including (+)-catechin (C), (-)-epicatechin (EC), (+)-gallocatechin (GC), (-)-epigallocatechin (EGC), (-)-epicatechin gallate (ECG), (-)-gallocatechin gallate (GCG) and (-)-epigallocatechin gallate (EGCG). They have antioxidative, anti-inflammatory, antimutagenic, and anticarcinogenic potential (*3-13*).

High Performance Liquid Chromatography (HPLC)

HPLC method is currently the most useful approach for the routine analysis of catechins. It can provide good separation, quantification and high sensitivity. Treutter (*23*) developed a post-column derivatization method of catechins and proanthocyanidins for their selective detection following analytical HPLC separation of crude plant extracts and beverages. Bailey *et al.* (*24*) studied the non-volatile, water soluble constituents of black tea by using reversed phase HPLC with a photodiode array detector and a linear gradient separation. This is a qualitative study of black tea liquor. Caffeine, theobromine, (-)-epicatechin gallate, (-)-epigallocatechin gallate, gallic acid, chlorogenic acid, etc. were identified by the spectral data and the retention times given by reference compounds. Sakata *et al.* (*25*) reported the quantitative analysis of (-)-epigallocatechin gallate (EGCG) in tea by HPLC with a C-18 reversed-phase column. EGCG was eluted within 20 min by using methanol-water-acetic acid (20:75:5) as an eluent. Tryptophan was used as an internal standard. Goto et al. (*26*) developed a simple and fast HPLC method for eight tea catechins and caffeine using an ODS column and a water-acetonitrile-phosphoric acid mobile phase system. The detection limit of this method was approximately 0.2ng for all nine compounds and the quantitation curves were linear between 2 ng to 2 μg. Dalluge *et al.* (*27*) studied a variety of stationary phase and elution conditions for the liquid chromatographic determination of six green tea catechins. Comparison of six reversed-phase columns and evaluation of elution conditions used for the separation indicate that deactivated stationary phases and the presence of acid in the mobile phase is essential for both the separation of the catechins and the efficient chromatography of these compounds. Lin *et al.* (*28*) also reported an isocratic

HPLC procedure for simultaneous determination of six catechins, gallic acid, and three methyxanthines in tea water extract. Cosmosil C18-MS packed column and a solvent mixture of methanol/doubly distilled water/formic acid (19.5:80.2:0.3. v/v/v) as mobile phase was used in this method. Lee *et al.* (*29*) reported a HPLC method with coulometric array detection for the determination of green and black tea catechins and theaflavins in biological fluids and tissues. The detection limits for catechins and theaflavins are from 5 to 10 ng/mL of saliva, plasma, or urine. Institute for Nutraceutical Advancement-Method Validation Program (INA-MVP) (*30*) has developed and validated a HPLC method for the determination of catechins, gallic acid and caffeine in green tea leaves and powdered extract. The HPLC chromatogram shown in Figure 1 was obtained by using INA-MVP method for green tea powdered extract 80% that provided a good separation for catechins and caffeine.

Mass Spectrometric (MS) Method

Mass spectrometric method can provide both molecular weight and structural information which is very important for identification of the components of tea and for characterization of synthetic analogues. Mass spectrometric methods including EIMS, FABMS, and LC/ESIMS have been surveyed as tools for the detection of catechins in extracts of green tea by Miketova *et al.* (*31*). The study showed that LC/ESIMS was an appropriate method for the direct analysis of crude extracts of green tea. Furthermore, Zeeb *et al.* (*32*) developed an LC/APCI-MS method for the separation, identification and direct microscale determination of twelve catechins in green and black tea infusions. Standard catechin mixtures and tea infusions were analyzed by LC/APCI-MS with detection of protonated molecular ions and characteristic fragment ions for each compound. Monitoring of the catechin-specific retro Diels-Alder fragment ion at m/z 139 throughout the chromatogram provided a unique fingerprint for catechin content in the sample. This study showed that LC/APCI-MS is a useful tool for providing routine separation and identification of catechins at femtomole to low-picomole levels without extraction or sample pretreatment.

High Performance Thin Layer Chromatography (HPTLC)

Zhu and Xiao (*33*) reported quantitative analysis of catechins and caffeine in green tea by HPTLC in 1991. The quantitative determination of the catechins and caffeine was performed by using an HPTLC system which consists of a CAMG TLC/HPTLC scanner with monochromator. The mobile

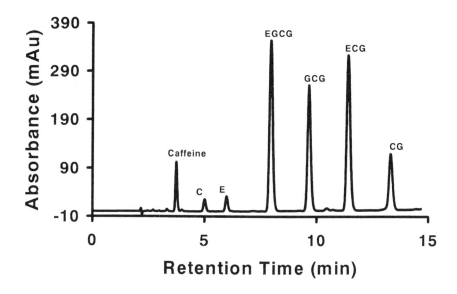

Figure 1. HPLC Chromatogram of Green Tea Leaf Powdered Extract 80% by Using INA-MVP Method

phase was chloroform/ethyl formate/*n*-butanol/formic acid (2:1.3:0.4:0.3). Quantitation was done by measuring the peak area at 254 nm with the TLC scanner.

Capillary Electrophoresis (CE)

Several catechins and caffeine can be separate by using HPLC. However, it is hard to measure ascorbic acid, theanine and other compounds simultaneously by HPLC method. Therefore, Horie *et al.* (*34*) developed simultaneous determination of qualitatively important components in green tea infusion using capillary electrophoresis. Separation was achieved using a fused-silica capillary column with a borax buffer at pH 8.0 and UV detection at 200 nm. The components analyzed were (-)-epicatechin, (-)-epigallocatechin, (-)-epicatechin gallate, (-)-epigallocatechin gallate, (+)-catechin, caffeine, theanine and ascorbic acid. However, only qualitation was done by this research. Arce *et al.* (*35*) published a further investigation. They reported a new method for the simultaneous determination of eleven polyphenol compounds in green tea infusions by flow injection-capillary electrophoresis in a single run. The components analyzed were caffeine, adenine, theophylline, epigallocatechin 3-gallate, epigallocatechin, epicatechin 3-gallate, (-)-epicatechin, (+)-catechin, gallic acid, quercetin and caffeic acid. Separation was achieved using a fused capillary column with 0.15 M borax buffer at a pH of 8.5, UV detection at 210 nm and 20 kV of voltage. Quantitative analysis was performed by the standard addition method.

Spectrophotometric Method

Spectrophotometric methods are commonly used for the analysis of total polyphenol compounds. However, Kivits *et al.* (*36*) developed a specific and sensitive colorimetric assay of total catechins from green and black tea in human biological fluids. Following precipitation of proteins with methanol, tea flavonoids were accumulated by solid phase adsorption to alumina. Reactions of adsorbed catechins with 4-dimethyl amino cinnamaldehyde under strong acidic conditions results in desorption and development of a highly selective colored complex. The method is highly specific for catechins with equimolar responses for catechin, epicatechin, epigallocatchin, epicatechin gallate and epigallocatechin gallate. Singh *et al.* (*37*) also reported that they synthesized a highly specific and sensitive diazotized sulfanilamide reagent for determination of tea catechins by spectrophotometric method at wavelength 425 nm.

Countercurrent Chromatographic Method

Countercurrent chromatography is a preparative all-liquid chromatographic technique based on partitioning of compounds between two immiscible liquid phases. It is used for preparative separation of natural compounds (*38*). Amarowicz and Shahidi (*39*) published a method for separation of individual catechins from green tea using a countercurrent chromatographic procedure. Solvent systems employed were water/chloroform (1:1) and water/ethyl acetate (1:1). Degenhardt *et al.* (*40*) applied high-speed countercurrent chromatography on a preparative scale to the separation of catechins, flavonol glycosides, proanthocyanidins, and strictinin from green and black tea.

Methods for the Analysis of Caffeine

Tea as a popular beverage has stimulating effect which is due to the presence of purine bases, including caffeine, theobromine and theophylline (*41,42*). Caffeine is the major alkaloid of tea, present in the range of 3-4% of the total weight (*43*).

Spectrophotometric Method

Spectrophotometric method used for determination of caffeine in tea was reported by Ullah *et al.* (*44*) and approved as a Chinese national standard method by the National Standard Bureau in China (*43*). The method involves in three extraction and filtration steps. The filtrate finally was analyzed at 274 nm by UV spectrophotometer.

Gas Chromatographic (GC) Method

GC method is not a comment method for the analysis of compounds like caffeine. However, Guo and Wan (*43*) developed a simple and quick GC method based on the sensitive response of a nitrogen-phosphorus detector to caffeine in solubility between caffeine and other polar compounds in green tea. The advantage of the GC method is its simplicity which saves considerable amounts of reagents and time.

High Performance Liquid Chromatographic (HPLC) Method

HPLC method is widely used for simultaneous analysis of tea polyphenols and caffeine which has been discussed in the previous section. Naik and Nagalakshmi (42) reported an improved HPLC method for the determination of only caffeine in tea products. The commonly used HPLC method for the analysis of caffeine content in tea employs direct injection of the samples in the column which would gradually reduces the efficiency of the column and shortens its life. In this modified method, a Sep-Pak C18 cartridge was used to remove the interfering tea pigments. Then, the sample was injected on a reverse phase μ-Bondapak C18 column employing acetonitrile and water (20:80 v/v) as mobile phase. More than 25 samples can be prepared using a single Sep-Pak C18 cartridge following the cleanup procedure carefully. The reverse phase μ-Bondapak C18 column resolution was good even after analysis of about 400 samples. A recovery of 98-100% caffeine was obtained from the Sep-Pak C18 cartridge.

Methods for the Analysis of Phenolic Pigments

Theaflavins and thearubigins are the phenolic pigments formed in the fermentation stage of black tea manufacture. They play an important role in the quality of black tea infusions, such as color and taste (2, 45).

Spectrophotometric Method

A spectrophotometric procedure for estimating theaflavins and thearubigins in black tea liquors in assessment of quality of teas was developed by Roberts and Smith (46) and modified by Ullah (47). After extraction by different solvent, the fractions of tea extract were measured at 380 and 460 nm. The 380 nm values are used to determine the theaflavins and the thearubigins by the equations.

Hilton in 1983 (48) developed a spectrophotometric method for theaflavin analysis which were modified by Robertson and Jewell (49). After water and other solvent extraction of tea, the final fraction was measured at 625nm. Theaflavin was calculated on a dry weight basis according to the formula. However, McDowell et al. (50) reported that flavonol glycosides were found to be present in the fraction used to determine the theflavins and the thearubigins. Therefore, theaflavins and thearubigins determined spectrophtometrically result in overestimation of these components due to flavonol glycosides interferes. In order to avoid the interference from flavonol glycosides, Whitehead and Temple (*51*) reported a spectrophotometric method for measuring thearubigins and theaflavins in black tea using C18 sorbent cartridges as separation tool.

Near-Infrared Reflectance Method

Hall *et al.* (*52*) developed a near-infrared spectroscopic method for the determination of theaflavin and the assessment of overall tea quality. Spectrophotometric method was used as reference method.

High Performance Liquid Chromatographic (HPLC) Method

Thearubigins are the name originally assigned to all the acidic brown pigments of black tea (*53*). A broad classification of these compounds is based on extractabilities by certain solvent (*54*). SI thearubigins is those extractable into ethyl acetate. Wedzicha and Donovan (*55*) investigated the potential for normal phase HPLC of derivatized SI thearubigins because of the high affinity of the thearubigins for the stationary phase. According to this method, SI thearubigins were extracted from black tea infusions and converted into acetyl and methyl derivatives which analyzed by silica HPLC columns using mixtures of chloroform and methanol as mobile phase. Bailey and Nursten (*56*) did a comparative study of the reversed-phase high-performance liquid chromato-graphy of black tea liquors with special reference to the thearubigins. Very good resolution of a wide range of phenolic tea pigments was obtained using a Hypersil ODS column with a citrate buffer, and eight theaflavins were observed in a single chromatogram for the first time. Bailey *et al.* (*57*) isolated a brown thearubigin fraction from a black tea liquor using chromatography on a column of Solka-Floc cellulose. They ran HPLC which showed the fraction to be a mixture of polymers. In 1994, Bailey and Nursten (*58*) had further research of isolation and high-performance liquid chromatographic analysis of thearubigin fractions from black tea.

Methods for the Analysis of Flavonol Glycosides

Flavonol glycosides are one of the most important groups of polyphenols in tea with particular physiological activity (*59,60*). Since flavonol glycosides are not affected by polyphenoloxidase as with catechins, their content in black tea is almost at the same level as in the green tea (*61*).

High Performance Liquid Chromatography (HPLC)

Hasler and Sticher in 1990 (*62*) reported HPLC method for the determination of five flavonoid aglycones. The method included hydrolyzing flavonol glycosides by using methanol and hydrochloric acid, and then using C18 reversed phase column with gradient separation. Finger *et al.* in 1991 (*61*) developed HPLC method for preparative isolation and analytical separation of kaempferol and quercetin rhamnodiglucosides which are the two characteristic flavonol glycosides in tea. Hertog *et al.* in 1993 (*63*) developed an RP-HPLC with UV detection for the determination of the content of quercetin, kaempferol, myricetin, apigenin and luteolin in twelve types of tea infusion. However, no luteolin and apigenin were detected. Price *et al.* in 1998 (*64*) investigated flavonol glycoside content and composition of tea infusions made from commercially available teas and tea products by HPLC. The analytical HPLC method which they developed involved gradient separation by using water-tetrahydrofuran-TFA (98/2/0.1) as mobile phase A and acetonitrile as mobile phase B. The column used was packed with Prodigy 5μ ODS (*3*) reversed phase silica and the detection wavelength is at 270 nm.

Methods for the Analysis of Volatile Flavor Compounds

Tea is one of the widely consumed beverages in the world because of its pleasant flavor aside from its other effects. Volatile compounds of tea have been investigated by many researchers using Gas Chromatography-Mass Spectrometry (GC-MS) (*65-70*) which is currently the most useful approach for the research of volatile tea flavor compounds combing with simultaneous distillation and extraction (SDE) or adsorptive concentration method, etc.

Conclusions

In conclusion, in order to control the quality and ensure the reliability and repeatability of pharmacological and clinical research, it is necessary to develop validated analytical methods for the qualitative and quantitative analysis of the active compounds found in tea products. HPLC, LC-MS, CE, UV/VIS, GC, GC-MS and NIR method have been developed for the analysis of the active components in tea products.

References

1. Mukhtar, H.; Wang, Z. Y.; Katiyar, S. K.; Agarwal, R. *Preventive Medicine* **1992**, *21*, 351-360.
2. Graham, H. N. *Preventive Medicine* **1992**, *21*, 334-350.
3. Lin, Y. L.; Juan, I. .; Chen, Y. L.; Liang, Y. C.; Lin, J. K. *J. Agric. Food Chem.* **1996**, *44*, 1387-1394.
4. Wiseman, S. A.; Balentine, D. A.; Frei, B. *Crit. Rev. Food Sci. Nutr.* **1997**, *37*, 705-718.
5. Conney, A. H.; Wang, Z. Y.; Huang, M. T.; Ho, C.-T.; Yang, C. S. *Preventive Medicine* **1992**, *21*, 361-369.
6. Hayatsu, H.; Inada, N.; Kakutani, T.; Arimoto, S.; Negishi, T.; Mori, K.; Okuda, T.; Sakata, I. *Preventive Medicine* **1992**, *21*, 370-376.
7. Stich, H. S. *Preventive Medicine* **1992**, *21*, 377-384.
8. Chen, J. *Preventive Medicine* **1992**, *21*, 385-391.
9. Bushman, J. L. *Nutrition and Cancer* **1998**, *31*, 151-159.
10. Yang, C. S.; Wang, Z. Y. *J. Natl. Cancer Inst.* **1993**, *58*, 1038-1049.
11. Dreosti, I. E.; Wargovich, M. J.; Yang, C. S. *Crit. Rev. Food Sci. Nutr.* **1997**, *37*, 761-770.
12. Hollman, P. C. H.; Tijburg, L. B. M.; Yang, C. S. *Crit. Rev. Food Sci. Nutr.* **1997**, *37*, 719-738.
13. Katiyar, S.; Mukhtar, H. *Int. J. Oncol.* **1996**, *57*, 78-83.
14. Engelhardt, U. H.; Finger, A.; Herzig, B.; Kuhr, S. *Dtsch. Lebensm. Rundsch.* **1992**, *88*, 69-73.
15. Kiehne, A.; Lakenbrink, C.; Engelhardt, U. H. *Z. Lebensm. Unters. Forsch. A* **1997**, *205*, 153-157.
16. Ribereau-Gayon, P. *Plant Phenolics*; University review series 3; Heywood, V. H., Ed.; Oliver and Boyd Press: Edinbergh, London, 1972; Chapter 2, pp 23-53.
17. Swain, T.; Goldstein, J. L. *Methods in Polyphenol Chemistry* . *Proceedings of the Plant Phenolics Group Symposium;* Oxford, April 1963; Pridham, J. B., Ed.; Pergamon Press: London, 1964; Chapter II, pp 131-146.
18. Robards, K.; Antolovich, M. *Analyst* **1997**, *122*, 11R-34R.
19. Pharmacopoeia Sinica 1995th Edition, Volume I, Appendix pp 30.
20. McClure, W. F. *Anal. Chem.* **1994**, *66*, 43A-53A.
21. Robert, P.; Bertrand, D.; Devaux, M. F.; Grappin, R. M. *Anal. Chem.* **1987**, *59*, 2187-2191.
22. Schulz, H.; Engelhardt, U. H.; Wegent, A.; Drews, H. H.; Lapczynski, S. *J. Agric. Food Chem.* **1999**, *47*, 5064-5067.
23. Treutter, D. *J. Chromatogr.* **1989**, *467*, 185-193.

226

24. Bailey, R. G.; McDowell, I.; Nursten, H. E. *J. Sci Food Agric.* **1990**, *52*, 509-525.
25. Sakata, I.; Ikeuchi, M.; Maruyama, I.; Okuda, T. *Yakugaku Zasshi* **1991**, *111*, 790-793.
26. Goyo, T.; Yoshida, Y.; Kiso, M.; Nagashima, H. *J. Chromatogr. A* **1996**, *749*, 295-299.
27. Dalluge, J. J.; Nelson, B. C.; Thomas, J. B.; Sander, L. C. *J. Chromatogr. A* **1998**, *793*, 265-274.
28. Lin, J. K.; Lin, C. L.; Liang, Y. C.; Lin-Shiau, S. Y.; Juan, I M. *J. Agric Food Chem.* **1998**, *46*, 3635-3642.
29. Lee, M. J.; Prabhu, S.; Meng, X.; Li, C.; Yang, C. S. *Anal. Biochem.* **2000**, *276*, 164-169.
30. Institute for Nutraceutical Advancement-Method Validation Program, Denver, CO, 2000.
31. Miketova, P.; Schram, K. H.; Whitney, J. L.; Kerns, E. H.; Valcic, S.; Timmermann, B. N.; Volk, K. J. *J. Nat. Prod.* **1998**, *61*, 461-467.
32. Zeeb, D. J.; Nelson, B. C.; Albert, K.; Dalluge, J. J. *Anal. Chem.* **2000**, *72*, 5020-5026.
33. Zhu, M.; Xiao, P. *Phytotherapy Research* **1991**, *5*, 239-240.
34. Horie, H.; Mukai, T.; Kohata, K. *J. Chromatogr. A* **1997**, *758*, 332-335.
35. Arce, L.; Rios, A.; Valcarcel, M. *J. Chromatogr. A* **1998**, *827*, 113-120.
36. Kivits, G. A. A.; van der Sman, F. J. P.; Tijberg, L. B. M. *Internat. J. Food Sci. Nutr.* **1997**, *48*, 387-392.
37. Singh, H. P.; Ravindranath, S. D.; Singh, C. *J. Agric. Food Chem.* **1999**, *47*, 1041-1045.
38. Marston, A.; Hostettmann, K. *J. Chromatogr. A* **1994**, *658*, 315-341.
39. Amarowicz, R.; Shahidi, F. *Food Res. Internat.* **1996**, *29*, 71-76.
40. Degenhardt, A.; Engelhardt, U. H.; Lakenbrink, C.; Winterhalter, P. *J. Agric. Food Chem.* **2000**, *48*, 3425-3430.
41. Rall, T. W. *The Pharmacological Basis of Therapeutics*, Gilman, A. G.; Goodman, L. S.; Gilman, A. Eds., Macmillan: New York, 1980, pp. 592-607.
42. Naik, J. P.; Nagalakshmi, S. *J. Agric. Food Chem.* **1997**, *45*, 3973-3975.
43. Guo, B.; Wan, H. *J. Chromatogr.* **1990**, *505*, 435-437.
44. Ullah, M. R.; Gogi, N.; Baruah, S. *Two Bud* **1987**, *34*, 50-53.
45. Balentine, D. A.; Wiseman, S. A.; Bouwens, L. C. M. *Crit. Rev. Food Sci. Nutr.* **1997**, *37*, 693-704.
46. Roberts, E. A. H.; Smith, R. F. *Analyst* **1961**, *86*, 94-98.
47. Ullah, M. R. *Curr. Sci.* **1972**, *41*, 422-423.
48. Hilton, P. J. *Encyclopaedia of Industrial Chemical Analysis*, John Wiley: New York, 1983, Vol. 18, 455-516.

49. Robertson, A.; Jewell, K. *Studies on Black Tea Chemistry.* Confidential Report, CFPRA, 1986.
50. McDowell, I.; Bailey, R. G.; Howard, G. *J. Sci. Food Agric.* **1990**, *53*, 411-414.
51. Whitehead, D. L.; Temple, C. M. *J. Sci. Food Agric.* **1992**, *58*, 149-152.
52. Hall, M. N.; Robertson, A.; Scotter, C. N. G. *Food Chem.* **1988**, *27*, 61-75.
53. Roberts, E. A. II. *J. Sci. Food Agric.* **1958**, *9*, 381-384.
54. Roberts, E. A. H.; Cartwright, R. A.; Oldschool, M. *J. Sci. Food Agric.* **1957**, *8*, 72-75.
55. Wedzicha, B. L.; Donovan, T. J. *J. Chromatogr.* **1989**, *478*, 217-224.
56. Bailey, R. G.; Nursten, H. E. *J. Chromatogr.* **1991**, *542*, 115-128.
57. Bailey, R. G.; Nursten, H. E.; McDowell, I. *J. Sci. Food Agric.* **1992**, *59*, 365-375.
58. Bailey, R. G.; Nursten, H. E. *J. Chromatogr. A* **1994**, *662*, 101-112.
59. Bertram, B. *Dtsch. Apoth. Ztg.* **1989**, *129*, 2561-2571.
60. Spilkova, J.; Hubik, J. *Pharm. Unserer. Zeit.* **1988**, *17*, 1-9.
61. Finger, A.; Engelhardt, U. H. *J. Sci. Food Agric.* **1991**, *55*, 313-321.
62. Hasler, A.; Sticher, O. *J. Chromatogr.* **1990**, *508*, 236-240.
63. Hertog, M. G.; Hollman, P. C. H.; van de Putte, B. *J. Agric. Food Chem.* **1993**, *41*, 1242-1246.
64. Price, K. R.; Rhodes, M. J. C.; Barnes, K. A. *J. Agric. Food Chem.* **1998**, *46*, 2517-2522.
65. Owuor, P. O. *J. Sci. Food Agric.* **1992**, *59*, 189-197.
66. Shimoda, M.; Shigematsu, H.; Shiratsuchi, H.; Osajima, Y. *J. Agric. Food Chem.* **1995**, *43*, 1616-1620.
67. Kawakami, M.; Ganguly, S. N.; Banerjee, J.; Kobayashi, A. K. *J. Agric. Food Chem.* **1995**, *43*, 200-207.
68. Shimoda, M.; Shigematsu, H.; Shiratsuchi, H.; Osajima, Y. *J. Agric. Food Chem.* **1995**, *43*, 1621-1625.
69. Kumazawa, K.; Masuda, H. *J. Agric. Food Chem.* **1999**, *47*, 5169-5172.
70. Wang, D.; Yoshimura, T.; Kubota, K.; Kobayashi, A. *J. Agric. Food Chem.* **2000**, *48*, 5411-5418.

Bioactivity of Nutraceuticals

Chapter 16

Antioxidants in Ginger Family

Nobuji Nakatani and Hiroe Kikuzaki

Food Chemistry, Graduate School of Human Life Science,
Osaka City University, Osaka 558–8585, Japan

Some species belonging to the Ginger family were assayed for antioxidant activity and their active components were characterized. From the rhizome of *Zingiber officinale*, the most common ginger, more than 50 compounds including gingerol-related compounds and diarylheptanoids were isolated. Their antioxidant assessment was performed by various assay systems and their chemical structure-activity relationship was overviewed. Twenty-six curcuminoids were isolated from *Curcuma domestica* and other *Curcuma* species. Some of them also showed potent activity to inhibit inflammation induced by TPA as well as antioxidant activity. The fruits of *Amomum tsao-ko*, one kind of cardamoms, included catechins as major active components.

Nutraceuticals or functional foods have been defined as foods containing compounds which provide human health benefits beyond the essential nutrients. Natural antioxidants are one of the most important and expected aspects of nutraceuticals. Antioxidants act not only to depress accumulation for oxidation products, such as lipid peroxides, but also to protect for oxidative damage in living cells which cause inflammation, cancer, atherosclerosis and aging. We have been engaged in the search for new natural antioxidants from spices and herbs, and in the determination of their structures and activity. Spices belonging to Labiatae, Zingiberaceae, Myristicaceae and Myrtaceae possess effective antioxidant activity. Herein, the chemistry and activity evaluation of the antioxidants in Ginger family are reviewed.

Antioxidants from the Zingiberaceae

Ginger (*Zingiber officinale*) is one of the most common spices. In the temperate and tropical Asian countries, many species of the Family Zingiberaceae are widely distributed and cultivated. Their rhizomes, fruits and seeds are used for spices, vegetables and traditional folk medicines (*1*). The family Zingiberaceae consists of 47 genera including 1400 species. Usage of typical genera such as *Alpinia, Amomum, Curcuma, Elettaria, Zingiber* are shown in Table I.

We examined the antioxidative activity for the acetone extracts of rhizomes of different species by the ferric thiocyanate (FTC) method and the thiobarbituric acid (TBA) method (*2,3*). As shown in Table II, most of the extracts depressed autoxidation of linoleic acid, showing stronger activity than or comparable to that of α-tocopherol, one of the natural antioxidants. Among them, common ginger (*Z. officinale*) showed the most remarkable activity and was as strong as butylated hydroxytoluene (BHT), a synthetic antioxidant, followed by turmeric (*Curcuma domestica* = *C. longa*), *Alpinia galanga*, *C. xanthorrhiza*, *Z. cassumunar*, *C. manga* and *A. kepulaga*.

Recently we screened 13 Malaysian species from the *Alpinia, Costus* and *Zingiber* genera for antioxidant activity (*4*). All dichloromethane and methanol extracts of their rhizomes showed higher activity than or comparable to that of α-tocopherol. In this screening study, the dichloromethane extracts of most species showed stronger activity than the methanol extracts, particularly observed in the extracts of *Costus megalobractea*, *C. spiralis* and *Z. cassumunar*. This observation suggests that less polar components present in the dichloromethane extracts contributed to the higher activity than did polar components in the methanol extracts.

Antioxidative Components of Zingiberaceae

Ginger (*Zingiber officinale*).

As one of the most important and popular spices, ginger has been cultivated in southern Asian countries for over 3,000 years. Ginger has been traditionally used in Chinese, Indian, Indonesian and Japanese medicines. Dichloromethane extract of the dried and powdered rhizomes of ginger was purified by a combination of chromatography of using silica gel, Sephadex LH-20, and octadecyl silica gel to afford more than 50 compounds, which were determined mainly by spectroscopic methods. These isolated compounds are

Table I . Usage of Typical Spices Belonging to the Family Zingiberaceae

Taxonomic name	English name	Parts used	Medicinal usage
Genus *Alpinia (Languas)*			
A. galanga	Galanga	Rhizome	Anorexia, Stomachache
		Flower	Skin disease
A. speciosa	Shell flower	Fruit	Stomachache
		Leaves	
Genus *Amomum*			
A. kepulaga	Round cardamom	Fruit	Cough, Gastritis
(A. cardamomum)		Rhizome	Tonic
A. subulatum	Black cardamom	Fruit	Stomachache
A. tsao-ko	Tsao-ko	Fruit	Stomachache
Genus *Curcuma*			
C. domestica	Turmeric	Rhizome	Diarrhea, Eczema
(C. longa)			Rheumatism
C. xanthorrhiza	Temu lawak (Java)	Rhizome	Anticonvulsant
			Anorexia, Malaria
Genus *Elettaria*			
E. cardamomum	Cardamom	Fruit	Stomachache
Genus *Zingiber*			
Z. cassumunar	Cassumunar ginger	Rhizome	Constipation, Obesity
			Anticonvulsant
Z. officinale	Ginger	Rhizome	Anorexia, Cold, Nausea
			Cough
Z. zerumbet	Zerumbet ginger	Rhizome	Abdominalgia, Anorexia
			Diarrhea

structually lassified into three groups, namely gingerol-related compounds, diarylheptanoids and cyclic diarylheptanoids as shown in Figure 1.

Table II. Antioxidant Activity of Rhizome of Zingiberaceae

Spice	FTC^a	TBA^b
BHT	99.0%	98.4%
α-tocopherol	86.8	90.1
Alpinia galanga	92.0	94.9
Amomum kepulaga	79.3	91.1
Curcuma aeruginosa	77.3	84.8
Curcuma domestica	96.3	95.4
Curcuma heyneana	63.1	74.9
Curcuma mangga	81.8	85.3
Curcuma xanthorrhiza	91.1	93.7
Phaeomeria speciosa	70.3	78.8
Zingiber cassumunar	84.4	91.1
Zingiber officinale	98.3	98.6

[a] Ferric thiocyanate method. [b] Thiobarbituric acid method.

Antioxidant activity was judged from inhibition of the absorbance at 1 day before (FTC, at 500 nm) or 1 day after (TBA, at 532 nm) the absorbance of control (absence of tested sample) reached a maximum. The results are expressed in % = [Absorbance of control–Absorbance in the presence of tested sample]/ Absorbance of control] × 100

Gingerol (**1**, systematically named as [6]-gingerol) is well known as one of the major pungent components. The structure of gingerol is a decane substituted with a 4-hydroxy-3-methoxyphenyl group at C-1, a carbonyl group at C-3 and a hydroxy group at C-5 with S configuration. Gingerol analogues have been found in ginger rhizome such as [3], [4], [7], [8], [10] and [12]-gingerol bearing 7, 8, 11, 12, 14, and 16 carbon-chain length, respectively (5-7).

Furthermore, demethoxy-gingerols and 4-O-methylated derivatives were also identified (6). Shogaol (**4**), a 4,5-dehydrated gingerol, is also known as a pungent component and a series of its related compounds such as [4], [6], [8] and [10]-shogaol were obtained. 3,5-Dihydroxy (**7**) and –diacetoxy (**8**) derivatives of gingerol were isolated in our survey research. Diarylheptanoids,

234

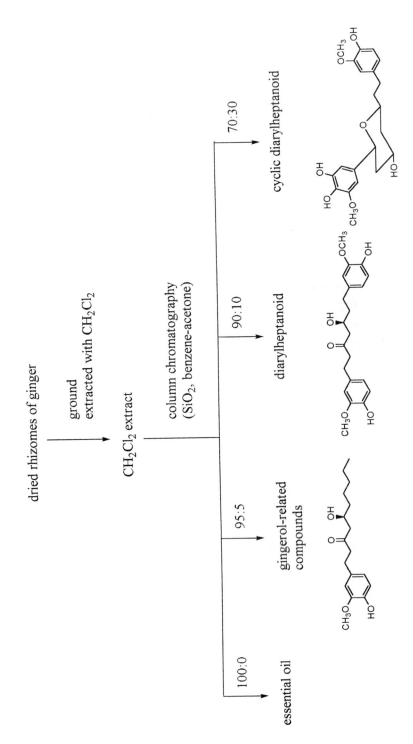

Figure 1. Extraction and purification of ginger (Zingiber officinale).

which are heptanes substituted with phenyl groups at C-1 and C-7, were obtained in smaller amount than the gingerols. The substitution patterns are similar to those of gingerol-related compounds. We have additionally found another type of diarylheptanoids, new cyclic diarylheptanoids (8). Totally more than 40 phenolic compounds were isolated as antioxidants from common ginger. As shown in Figure 2, these compounds are structurally classified into three types, which are further divided based on the substitution system. Concerning the antioxidant activity, the structure-activity relationships of isolated compounds were investigated (7). Through a measurement of the antioxidant activity among gingerol homologues (1-(4-hydroxy-3-methoxyphenyl)-5-hydroxy-alkan-3-one) with different alkyl chain (C_8, C_{10}, C_{12}, C_{14}), higher activity was observed in gingerols with longer length chains. Shogaol homologues showed similar results as gingerols. When compared the activity of [6]-gingerol-related compounds [1-(4-hydroxy-3-methoxyphenyl)decane] with different substituents at C-3~C-5 at a concentration of 100 μM of each sample in FTC measurement, [6]-gingerdiacetate showed the highest activity, followed by [6]-gingerdiol (7), ≧ [6]-shogaol (4)>[6]-gingerol (8)>[6]-dehydrogingerdione (10)>α-tocopherol. Similar tendency was observed in the activity efficiency of diarylheptanoids. When focused on the substitution pattern in benzene ring in both the case of gingerol-type homologues and diarylheptanoids, a 4-hydroxy-3-methoxyphenyl group contributed to the activity much more than a 4-hydroxy-3,5-dimethoxyphenyl or 3,4-dihydroxyphenyl group. These results suggest that the structural importance on affecting antioxidant activity are 1) the alkyl chain length, 2) the substitution pattern at alkyl chain and 3) the substitution pattern in benzene ring in both gingerol-related compounds and diarylheptanoids.

Further detailed antioxidant property of three gingerol-related components, [6]-gingerol (1), [6]-shogaol (4) and [6]-dehydrogingerdione (10), were evaluated based on following four measurements as well as the FTC measurement; 1) DPPH radical-scavenging activity, 2) the inhibitory effect on oxidation of methyl linoleate under aeration and heating using Oil Stability Index (OSI) method (9), 3) the inhibitory effect on oxidation of liposome induced by AAPH or AMVN, and 4) the inhibitory effect on AAPH-induced LDL peroxidation in human plasma. DPPH radical scavenging activity was in the order [6]-shogaol (4) > [6]-gingerol (1) > [6]-dehydrogingerdione (10), corresponding to the result from the FTC method.

On the other hand, the antioxidant effect based on the OSI method was in the order [6]-dehydrogingerdione (10) > [6]-shogaol (4) ≒ [6]-gingerol (1). These three compounds showed antioxidant effects on both AAPH-induced and AMVN-induced oxidation of liposome prepared from egg yolk phosphatidylcholine. In the case of AAPH-induced oxidation, these compounds

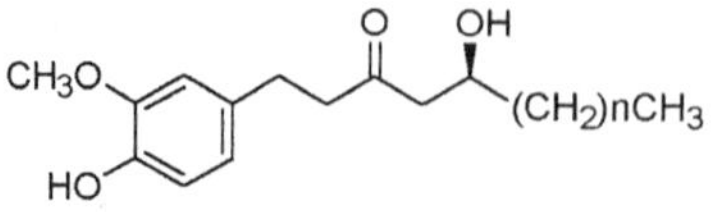

236

1: [6]-gingerol (n=4)
2: [8]-gingerol (n=6)
3: [10]-gingerol (n=8)

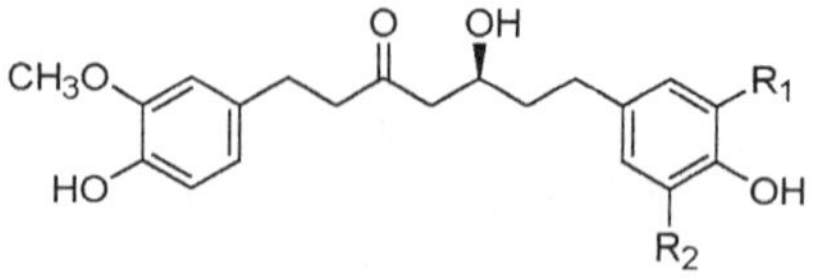

11: R_1=OCH$_3$, R_2=H (hexahydrocurcumin)
12: R_1=R_2=OCH$_3$
13: R_1=R_2=H

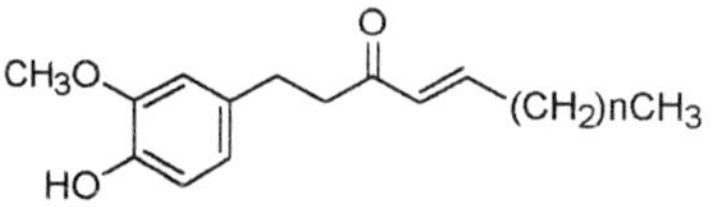

4: [6]-shogaol (n=4)
5: [8]-shogaol (n=6)
6: [10]-shogaol (n=8)

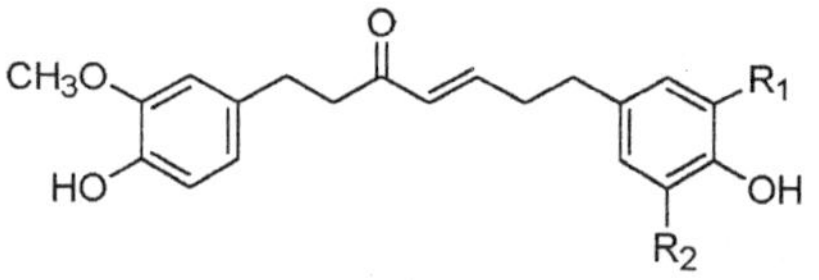

14: R_1=OCH$_3$, R_2=H (gingerenone A)
15: R_1=R_2=OCH$_3$
16: R_1=R_2=H

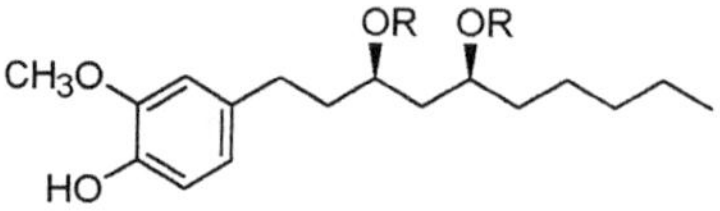

7: R=H ([6]-gingerdiol)
8: R=Ac ([6]-gingerdiacetate)

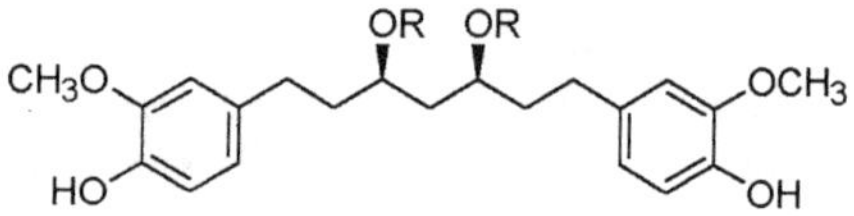

17: R=H (octahydrocurcumin)
18: R=Ac (3,5-diacetyloctahydrocurcumin)

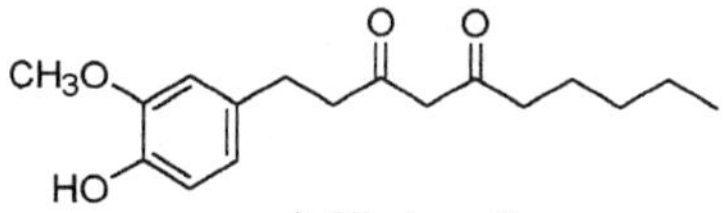

9: [6]-gingerdione

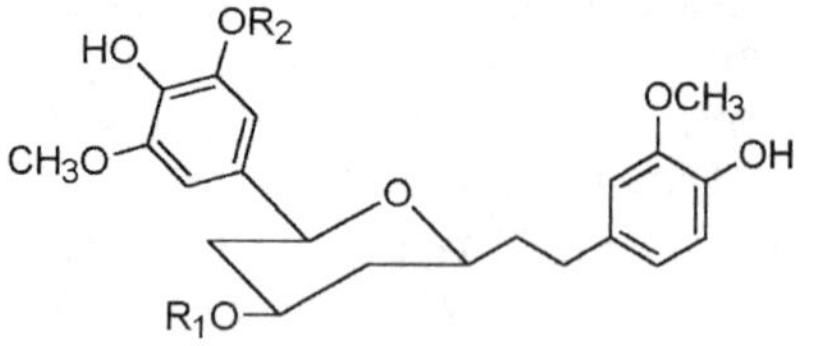

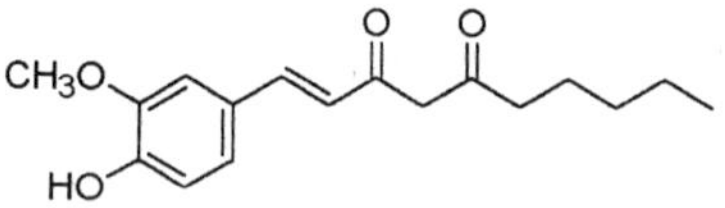

10: [6]-dehydrogingerdione

19: R_1=Ac, R_2=H
20: R_1=R_2=H
21: R_1=H, R_2=CH$_3$

Figure 2. Antioxidants isolated from ginger (Zingiber officinale).

depressed about 40% degree of oxidation and the tendency of activity was (**4**) > (**1**) > (**10**). When induced by AMVN (2,2'-azobis(2,4-dimethylvaleronitrile), lipid-soluble) , every gingerol related compound showed about 20% depression. On the AAPH (2,2'-azobis(2-amidinopropane)dihydrochloride, water-soluble)-induced LDL peroxidation, [6]-shogaol exhibited more effectiveness than two other compounds.

Turmeric (*Curcuma domestica* = *C. longa*).

Turmeric has been used for centuries through out India and southern Asia for its attractive yellow color and flavor. Dried and powdered rhizome is not only for a culinary spice of a great variety of dishes but also for medicinal purpose and dying textile. The yellow pigments, curcumin (**22**) and two demethoxylated curcumins (**23, 24**), are known to posses antioxitant activity (Figure 3). Further purification of turmeric, two additional curcumin related compounds (**25, 27**) (*10*) and two new diarylpentanoids (**28, 29**) (*11*) were determined. These curcuminoids showed strong antioxidant activity against autoxidation of linoleic acid, and anti-inflammatory activity measured on mouse ears by using a tumor promoter, TPA (12-*O*-tetradecanoylphorbol-13-acetate) as an inducer (*11*).

Tume lawak (*Curcuma xanthorrhiza*).

This species is also used as a spice and a traditional medicine in tropical Asian countries. A new antioxidant (**26**) was isolated together with 4 known curcuminoids from the dichloromethane extract (*12*).

Cassumunar ginger (*Zingiber cassumunar*).

The rhizome is cultivated for food and medicinal use in tropics. From the slightly polar fraction of acetone extract of fresh rhizomes, 6 new complex curcuminoid-phenylbutanoids were isolated and determined, being named cassumunin A~C (**30-32**) and cassumunarin A~C (**33-35**) (*13~15*). Cassumunarin A~C might be derived from curcumin by Diels Alder reaction with phenyl-butadiene which were also found in the same rhizome (*16*). The antioxidant activity of all these compounds was stronger than those of curcumin (**22**) and α-tocopherol, and their anti-inflammatory activity, measured by the TPA-induced method, is also higher than that of curcumin.

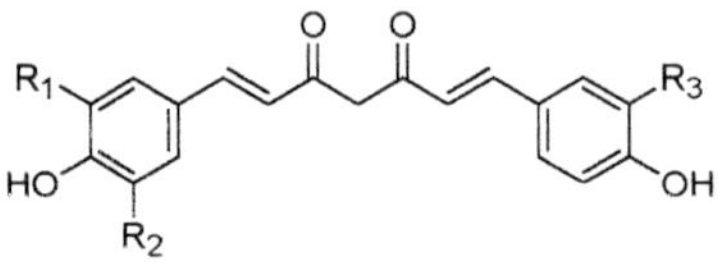

22: $R_1=R_3=OCH_3$, $R_2=H$ (curcumin)
23: $R_1=OCH_3$, $R_2=R_3=H$
24: $R_1=R_2=R_3=H$
25: $R_1=OCH_3$, $R_2=H$, $R_3=OH$
26: $R_1=R_2=R_3=OCH_3$

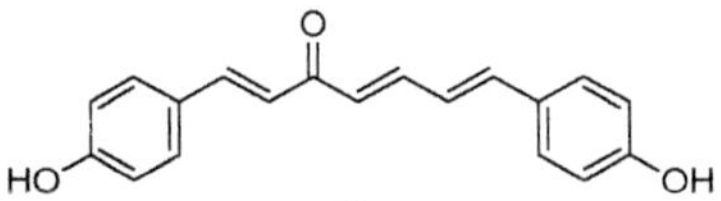

27

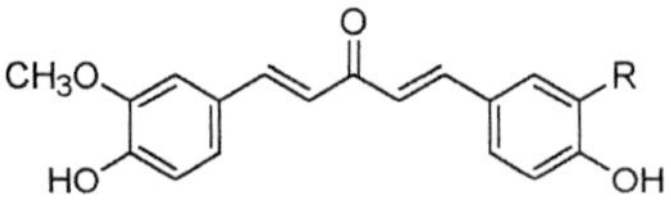

28: R=OCH_3
29: R=H

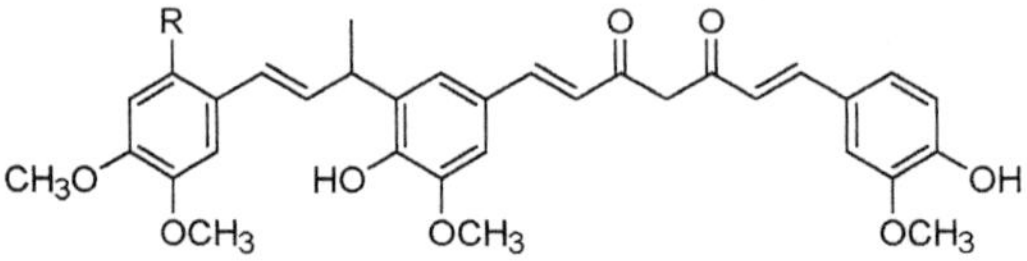

30: R=H (cassumunin A)
31: R=OCH_3 (cassumunin B)

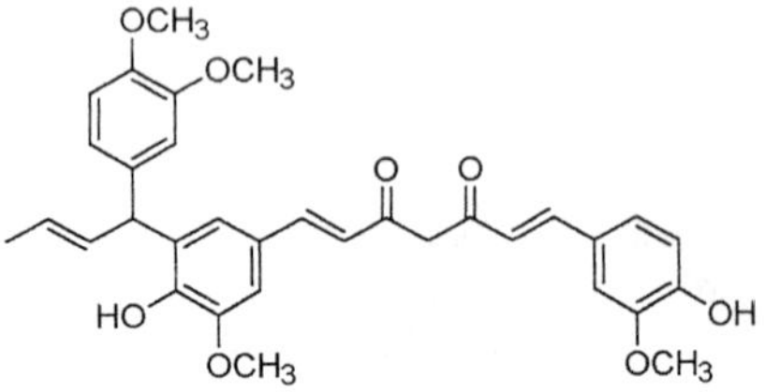

32: cassumunin C

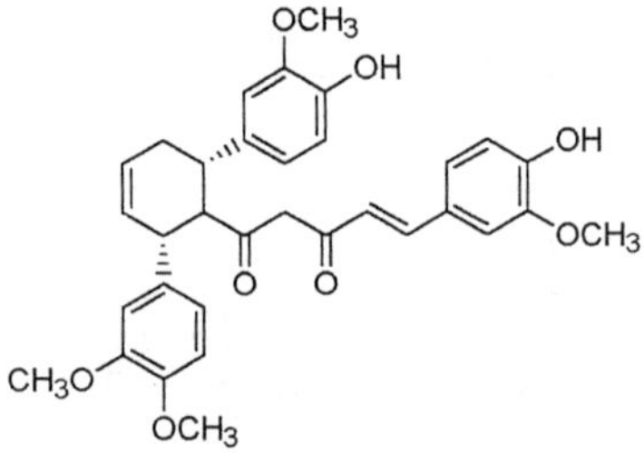

33: cassumunarin A

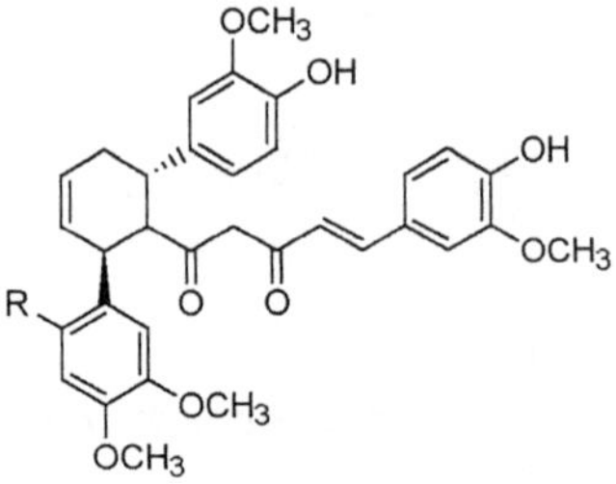

34: R=H (cassumunarin B)
35: R=OCH_3 (cassumunarin C)

Figure 3. Curcuminoid isolated from Zingiberaceae.

Tsao-ko (*Amomum tsao-ko*).

The fruits of *Amomum tsao-ko* is called tsao-ko, which has been used as a spice and a medicine for the treatment of stomach disorders, liver abscess, and infection of the throat. Preliminary screening of the antioxidant activity promoted us to engage in structure determination of active constituents. Eleven compounds were isolated from the ethyl acetate-soluble fraction of a 70% aqueous acetone extract. Among them, (+)-epicatechin, (-)-cathechin and catechol derivatives showed strong activities in DPPH radical scavenging activity measured by colorimetric and ESR, and in antioxidant activity measured by OSI method (*16*).

Conclusion

Antioxidant activity is one of the most important biological properties of secondary metabolites. There have been determined to be effective antioxidants in many different species of Zingiberaceae. In addition to the above-mentioned species, searching for effective components from other species such as *Alpinia galanga* and *A. speciosa* has been continuously going on. Furthermore, anti-oxidative compounds like curcuminoiods play an important role in the preventive effect against inflammation and tumor promotion step.

Acknowledgement

This work was supported by the Program for Promotion of Basic Research Activities for Innovative Biosciences (BRAIN).

References

1. Kikuzaki, H. In *Herbs, Botanicals & Teas;* Mazza. G.; Oomah, B. D., Eds.; Technomic Publishing: Lancaster, PA, 2000; pp 75-105.
2. Jitoe, A.; Masuda, T.; Tengah, I. G. P.; Suprapta, D. N.; Gara, I. W.; Nakatani, N. *J. Agric. Food Chem.* **1992,** *40*, 1337-1340.
3. Kikuzaki, H.; Nakatani, N. *J. Food Sci.* **1993,** *58*, 1407-1410.
4. Habsah, M.; Amran, M.; Mackeen, M. M.; Lajis, N. H.; Kikuzaki, H.; Nakatani, N.; Rahman, A. A.; Ghafar; Ali, A. M. *J. Ethnopharmacology,* **2000,** *72*, 403-410.

5. Masada, Y.; Inoue, T.; Hashimoto, K.; Fujioka, M.; Shirak, K. *Yakugaku Zasshi* **1973**, *93*, 318-321.
6. Harvey, D. J. *J. Chromatography* **1981**, *212*, 75-84.
7. Kikuzaki, H.; Kawasaki, Y.; Nakatani, N. In *Food Phytochemicals for Cancer Prevention II-Teas, Spices, and Herbs*, ACS Symposium Series 547; Ho, C.-T.; Osawa, T.; Huang, M-T.; Rosen, R. T., Eds.; American Chemical Society: Washington, D. C., 1994; pp 237-243.
8. Kikuzaki, H.; Nakatani, N. *Phytochemistry*, **1996**, *43*, 273-277.
9. Nakatani, N.; Tachibana, Y.; Kikuzaki, H. *J. Am. Oil Chem. Soc.* **2001**, *78*, 19-23.
10. Nakayama, R.; Tamura, Y.; Yamanaka, H.; Kikuzaki, H.; Nakatani, N. *Phytochemistry* **1993**, *33*, 501-502.
11. Masuda, T.; Jitoe, A.; Isobe, J.; Nakatani, N.; Yonemori, S. *Phytochemistry* **1993**, *32*, 1557-1560.
12. Masuda, T.; Isobe, J.; Jitoe, A.; Nakatani, N. *Phytochemistry* **1992**, *31*, 3645-3647.
13. Masuda, T.; Jitoe, A.; Nakatani, N. *Chem. Lett.* **1993**, 189-192.
14. Masuda, T.; Jitoe, A. *J. Agric. Food Chem.* **1994**, *42*, 1850-1856.
15. Masuda, T.; Jitoe, A.; Mabry, T. J. *J. Am. Oil Chem. Soc.* **1995**, *72*, 1053-1057.
16. Jitoe, A.; Masuda, T.; Nakatani, N. *Phytochemistry* **1993**, *32*, 357-363.
17. Martin, T. S.; Kikuzaki, H.; Hisamoto, M.; Nakatani, N. *J. Am. Oil Chem. Soc.* **2000**, *77*, 667-673.

Chapter 17

Dietary Phytoestrogens: Safety, Nutritional Quality, and Health Considerations

G. Sarwar, M. R. L'Abbé, S. P. J. Brooks, E. Lok, G. M. Cooke, O. Pulido, and P. Thibert

Food Directorate, Health Products and Food Branch, Health Canada, Banting Research Building (AL: 2203C), Tunney's Pasture, Ottawa, Ontario K1A OL2, Canada

Possible health benefits and potential adverse effects of dietary phytoestrogens are discussed. Moreover, data from a 16-week rat study to investigate the effects of the addition of graded levels (0, 50, 100, 200 and 400 mg/kg diet) of soybean isoflavones to a casein control diet on growth, plasma total cholesterol, plasma isoflavones and length of estrus cycle are reported. An isoflavones-rich extract (Novasoy) and a soy infant formula were used a source of dietary isoflavones. Initial analyses revealed a lower body weight in male rats fed the highest two levels of isoflavones. There was a dose-related increase in the levels of plasma isoflavones. The plasma data showed that the absorption of isoflavones from formula was lower than that from Novasoy. There was also a dose-related increase in the length of the estrus cycle in female rats. The cycle in rats fed the formula diet was, however, shorter than that in those fed the Novasoy diet providing the same amount of isoflavones. This suggested differences in the potency based on the source of isoflavones.

Isoflavones and other phytoestrogens (coumestans and lignans) are estrogenic compounds found in plants (*1*). They exert estrogenic effects on the central nervous system, induce estrus and stimulate growth of the genital tract of female animals. Phytoestrogens can be divided into three main classes such as isoflavones, coumestans and lignans. They all are diphenolic compounds with structural similarities to natural and synthetic estrogens and antiestrogens. Isoflavones and other phytoestrogens exhibit weak estrogenic activity. The antiestrogenic activity of phytoestrogens may be partially explained by their competition with endogenous estradiol for binding to nucleus. This raises the possibility that they may be protective in hormone-related diseases such as breast cancer and other cancers as evidenced by epidemiological evidence (*2-6*). Isoflavones also have antioxidant effects that can potentially reduce heart disease.

Animal studies have demonstrated adverse effects of high intake of dietary isoflavones and other phytoestrogens on fertility and interference with sexual differentiation of the brain and reproductive development (*4*) and thyroid peroxidase activity (*7*). There is a lack of information regarding potential harmful effects of high intakes of phytoestrogens in humans, especially in infants fed soy-based formulas. The daily exposure of infants to isoflavones in soy-based formulas was calculated to be 6- to 11-fold higher on a weight basis than the dose that has hormonal effects in adults consuming soy foods. However, the No Observed Adverse Effect Level (NOAEL) of dietary isoflavones is not known.

A 16-week rat feeding study was conducted at Health Canada to assess the safety and nutritional quality of soy isoflavones. In this study, the biological effects of the addition of graded levels (0, 50, 100, 200 and 400 mg/kg diet) of soybean isoflavones to a casein control diet in male and female rats were investigated. The objectives of this manuscript are to review possible beneficial effects and potential adverse effects of dietary phytoestrogens, and to report the preliminary data from the Health Canada study on safety assessment of soy isoflavones.

Food Sources of Phytoestrogens

Phytoestrogens are present in many foods including beans, sprouts, cabbage, spinach, soybeans and grains. Soybeans, soy-protein products and soy-based infant formulas are the most significant sources of isoflavones (a major subclass of phytoestrogens) and flax is the richest source of lignans (*8*). Soy isoflavones are now marketed as powders, and tablets may soon be available over the counter. Isoflavones occur naturally in plants either as glycoside conjugates, or in unconjugated form. In seeds, isoflavones are present mainly as glucosides which are biologically inert. On ingestion, these glucosides are readily hydrolysed in the acidic environment of the stomach and by the intestinal bacteria to release the unconjugated isoflavones which are biologically active. Consequently, isoflavones are rapidly and efficiently digested and absorbed with plasma concentrations rising significantly after ingestion of soy products. Elimination of isoflavones occurs largely via the urine, mainly as glucoronide conjugates, sharing the physiological features and behaviour of endogenous estrogens

The isoflavones composition of soybeans and soy foods has been reviewed (9-11). In soybean seeds, the amount (1.6-4.2 mg daidzein + genistein/g dry basis) varies depending upon cultivars, crop year, and location of growth, etc. There is commercial interest in increasing the isoflavone content of soybeans through biotechnology. Soy flour, made by defatting and dehulling flakes should have an isoflavone profile approximating that of soybeans. Protein isolates contain reduced levels (0.6-1.0 mg/g dry basis) of isoflavones compared to soybean seeds and flours, as a result of processing during manufacturing. Values in protein concentrates depend upon the method of preparation; a concentrate made by aqueous alcohol extraction is low in isoflavones because the isoflavones are soluble in aqueous alcohol and thus largely removed during processing. Traditional nonfermented soy foods such as tofu and soy milk contain higher levels (2.0-2.1 mg/g dry basis) of total isoflavones compared to fermented products such as tempeh and miso (1.0-1.6 mg/g dry basis). Soy-based infant formulas sold in Canada contained 0.2-0.3 mg/g of total isoflavones (dry basis) (12). Novasoy, a commercial isoflavone-rich product prepared by Archer Daniel Midland contained 300 mg/g of daidzein and genistein. The United States Department of Agriculture (USDA) and Iowa State University have developed the first database on isoflavones collected from published reports. In addition, data on isoflavones were gathered at Iowa State by extensive sampling and laboratory analysis. This information has ben placed on USDA website, and can be accessed at their website: www.nal.usda.gov/fnic/foodcomp/isoflav.

Human Dietary Phytoestrogens Intake

Current information about the amount of phytoestrogens in the human diet is limited. Intake will depend on the type and amount of food consumed. For example, British isoflavone intake was estimated to be less than 1mg/d (13) but intakes in Asian countries such as Japan, Taiwan, Korea and China are significantly higher (50-100 mg/d; 10, 14-15). Infants fed soy-based formulas also receive

244

significant amounts (28-47 mg/d) of isoflavones (*16*). The potential use of soybean protein products in the USA is expected to increase significantly following an FDA (Food and Drug Administration) decision that will allow food manufacturers to make health claims for soy protein products as well as a USDA rule change that will allow suitable alternate protein products (such as soybean products) to replace 100% of the meat products in the National School Lunch Program, the Summer Food Service Program, and the Child and Adult Care Food Program. Prior to this change, alternate protein products could replace no more than 30% of meat products.

Beneficial Effects of Dietary Phytoestrogens

Many potential benefits of dietary phytoestrogens have been identified (*2-6*). These may include reduction of hormone-based cancer (breast and prostate) as well as epithelial cancers (endometrium, colon, rectum, stomach and lung). In addition, some authors have linked phytoestrogens to reduced incidence of osteoporosis and cardiovascular disease (*17*). In the case of the potential to reduce the risk of cancer, evidence is based on (not entirely consistent) epidemiological studies (*18*) and on comparisons of the rates of cancer across different countries where the populations consume different diets. For cancer of the prostate, colon, rectum, stomach and lung, the epidemiological evidence is most consistent for a protective effect resulting from a high intake of plant foods including grains, legumes, fruits and vegetables. It is not possible, however, to identify specific food components that may be responsible.

Dietary intervention studies have shown increased plasma or urine concentrations of phytoestrogens in women fed soy and flaxseed. These increased concentrations were observed in populations having a lower cancer incidence (*19*) suggesting a potential biochemical mechanism for the action of phytoestrogens. Thus, plausible biochemical mechanisms exist to explain the effect of soy-associated components. However, it is uncertain whether these biochemical changes actually mediate a measurable physiological effect. This is emphasized by a recent international symposium on the role of soy in preventing and treating chronic disease that concluded: "It is clear from the results presented and the discussion that took place during the symposium that with few exceptions, considerably more research is required before a good understanding of the health effects of soy can be realized." (*20*).

Soy may also help to reduce the risk of CVD and atherosclerosis through beneficial effects on blood lipids and cholesterol. In 1999, the FDA reviewed the evidence surrounding a health claim on the association between soy protein and reduced risk of heart disease. The FDA concluded that "soy protein included in a diet low in saturated fat and cholesterol may reduce the risk of coronary heart disease by lowering blood cholesterol levels." (21).

Other food components in addition to phytoestrogens may contribute to an overall protective effect; examples may include glucosinolates, indoles and possibly phytolexins in cruciferous vegetables and flavonoids (other than isoflavones) in fruits, vegetables and beverages such as tea and wine.

Conclusions

There is convincing epidemiological evidence linking a high intake of plant food with reduced risk of many hormone-dependent chronic diseases, including cancer. However, the evidence linking phytoestrogens to these chronic diseases is insufficient. Thus with the present state of knowledge, it is imprudent to recommend particular dietary practices or changes with respect to phytoestrogens consumption.

Adverse Effects of Dietary Phytoestrogens

1. Fertility: Ingestion of subterranean clover which contains high levels of isoflavones was reported to result in a number of reproductive problems in sheep including ovarian cysts, irreversible endometriosis, and a failure to conceive (22) . This condition, subsequently named clover disease, was reported to be caused by equol formed by the bacterial metabolism of formononetin, an isoflavone present in clover (23). Dietary phytoestrogens have also been implicated as playing a role in the reproductive failure and liver disease of captive cheetahs in North America (24) which were associated with the presence of large amounts of daidzein and genistein in their diet.

2. Early Development: To date, the study of the effects of phytoestrogens on fetal development has been limited to rats, and has demonstrated mainly adverse effects on sexual differentiation of the brain, maturation of neuroendocrine control of ovulation and puberty and development of the female reproductive tract (25).

3. A dose range study was conducted by feeding purified genistein to rats at 0-1250 ppm in a soy-free diet, from gestation to postnatal d 50 at the FDA NCTR (National Center for Toxicological Research) in Jefferson, Arkansas (7). Adverse effects included: (a) hypertrophy and hyperplasia of the mammary ducts and acini in males at 25 ppm and in females at 625 ppm, (b) hypospermia at the head of the epididymis, inflammation of the dorsal prostate and asynchronous cycles of the uterus and vagina at 625 ppm, and (c) degeneration of the ovaries and seminiferous tubules at 1250 ppm. There was also a dose-dependent decrease in thyroid

peroxidase (TPO) activity at 25-1250 ppm. The inhibition of TPO, which produces the thyroid hormone T3 and T4, can be expected to generate thyroid abnormalities.

4. Research conducted at FDA's NCTR also found that the metabolism of dietary isoflavones is significantly affected by sex (26). In this study, tissue distribution of genistein in rats was determined. The rats were exposed to genistein in utero, through maternal milk, and as adults through postnatal d 140 via essentially isoflavones-free feed fortified at 5, 100 and 500 µg/g with genistein aglycone. Pharmacokinetic analysis of serum genistein showed a significant difference ($P < 0.001$) in the elimination half life and area under the concentration-time curve between male (2.97 h and 22.3 µmol/(L·h), respectively) and female rats (4.26 h and 45.6 µmol/(L·h). Endocrine-responsive tissues including brain, liver, mammary, ovary , prostate, testis, thyroid and uterus showed significant dose-dependent increases in total genistein concentrations. Female liver contained the highest amount of genistein (7.3 pmol/mg tissue) and male whole brains contained the least (0.04 pmol/mg). The physiologically active aglycone form was present in tissues at fractions up to 100%, and the concentration was always greater than that observed in serum in which conjugated forms predominated (95-99%).

5. Genistein, a dietary estrogen, inhibits the growth of breast cancer cells at high doses but additional studies have suggested that at low doses, genistein stimulates proliferation of breast cancer cells (27). Genistein and the fungal toxin zearalenone were found to increase the activity of cyclin dependent kinase 2 (Cdk2) and cyclin D_1 synthesis and stimulate the hyperphosphorylation of the retinoblastoma susceptibility gene product pRb105 in human breast cancer cells. Dietary estrogens not only failed to suppress DDT-induced Cdk2 activity, but were found to slightly increase enzyme activity. It was concluded that dietary estrogens at low concentrations do not act as antiestrogens, but act like DDT and estradiol to stimulate human breast cancer cells to enter the cell cycle.

6. To examine association of midlife tofu consumption with brain function and structural changes later in life, White et al. (28) studied surviving participants (all Japanese Americans living in Hawaii) of a longitudinal study established in 1965 for research on hear disease, stroke and cancer. Information on consumption of selected foods was available from standardized interviews conducted during 1965-1967 and 1971-1974. It was found that men who consumed greater amounts of tofu during midlife appeared to score worse on cognitive tests, to have lower brain weight and to demonstrate ventricular enlargement on MRI compared to men with lower tofu intake. For example, cognitive impairment was identified in 4% of men with the lowest tofu intake, compared to 19% of those with the highest intake; low brain weight was observed in 12% of subjects with the lowest tofu intake and 40% of men in the highest category. Supporting evidence was provided by apparent dose-response effects (i.e. increasing risk of a poor outcome with increasing levels of tofu consumption) and by the subjects' wives (logically assumed to have comparable tofu consumption to their husbands), in whom a similar association

between tofu and lower cognitive scores was noted. These preliminary results emphasize the need for large population-based studies of diet and cognitive functions.

7. Safety concerns in infants and children: Because phytoestrogens possess a wide range of hormonal and non-hormonal activities, it has been suggested that adverse effects may occur in infants fed soy-based infant formulas. Available evidence suggests that infants can digest and absorb dietary isoflavones in active forms. Since neonates are generally more susceptible than adults to perturbations of the sex steroid milieu, it would be highly desirable to study the effects of soy isoflavones on steroid-dependent developmental processes in human babies (29). Concentrations of plasma isoflavones in infants fed soy formula were 13,000 to 22,000 times higher than plasma estradiol levels in early life and may have been sufficient to exert biological effects (16). On the other hand, the contribution of isoflavones from breast-milk and cow-milk is negligible suggesting further research to assess the biological effects of phytoestrogens exposure early in life (16).

It has been suggested that feeding practices in infancy may affect the development of various autoimmune diseases later in life. Since thyroid alterations are among the most frequently encountered autoimmune conditions in children, Fort et al. (30) studied whether breast and soy-based formula feeding in early life were associated with the subsequent development of autoimmune thyroid disease. A detailed history of feeding practices was obtained in 59 children with autoimmune thyroid disease, their 76 healthy siblings, and 54 healthy nonrelated control children. There was no difference in the frequency and duration of breast feeding in early life among the three groups of children. However, the frequency of feedings with soy-containing formulas in early life was significantly higher in children with autoimmune thyroid disease (prevalence 31%) as compared with their siblings (prevalence 13%). This retrospective analysis documented the association of soy formula feeding in early infancy and autoimmune thyroid disease.

Health Canada Study to Assess Safety of Soybean Isoflavones in the Rat Model

A 16-weeks rat feeding study to investigate the biological effects of the addition of graded levels (0, 50, 100, 200 and 400 mg/kg diet) of soybean isoflavones to a casein control diet, adequate in all nutrients for rat growth (31) has been completed in the Food Directorate Laboratories of Health Canada. An isoflavones-rich extract (Novasoy) and a soy-based infant formula were used as the source of dietary isoflavones. Six semi-purified diets (casein containing zero isoflavones, casein + 50 mg/kg diet of isoflavones from Novasoy, casein + 100 mg/kg diet of isoflavones from Novasoy, casein + 200 mg/kg diet of isoflavones from Novasoy, casein + 400 mg/kg diet of isoflavones from Novasoy, and casein

+ 200 mg/kg diet of isoflavones from a soy-based infant formula) were tested in both male (ten per diet) and female (ten per diet) rats. Isoflavones (daidzin, daidzein, genistin and genistein) in foods were determined by the method of Wang and Murphy (*32*) using Waters HPLC linear gradient with UV detector monitored at 254 nm, while isoflavones and their metabolites (daidzein, genistein, equol and 4-ethyl phenol) in plasma were determined by the method of King et al. (*28*) with the following exceptions: Waters NOVA-PAK C-18 15 cm column, Waters 464 electrochemical detector, 0.75 volt potential, and mobile phase consisting of 40:50:1 v/v/v: methanol: 0.1 mol/L ammonium acetate pH 4.6: 25 mmol/L EDTA (instead of 50:50:1 used by King et al.) (*33*). As shown in Figure 1, this method allowed clear separation of daidzein, genistein, equol and 4-ethyl phenol in blood samples in about 18 minutes. Levels of plasma total cholesterol were determined according to the high performance gel filtration chromatographic method of Kieft et al. (*34*). In female rats, vaginal smears were collected with saline-moistened Q-tips and spread on pre-cleaned labelled slides. The smears were immediately sprayed and coated with a water soluble fixative. Collected slides were later stained with Wright-Giemsa Stain. The slides were allowed to airdry and cover-slipped with mounting medium. The stained slides were examined with the aid of a high-power microscope to determine the stage of the estrus cycle of the rat according to the types of vaginal cells (epithelial, cornified and leukocytes) observed. The proestrus, estrus, metestrus-I, metestrus-II and diestrus stage of the estrus cycle was determined according to the types and quantity of cells present. The length of the estrus cycle was determined by the number of estrus observed during the 14-day period expressed as days between estrus.

Preliminary results of this study on growth, levels of plasma total cholesterol, levels of plasma isoflavones and length of estrus cycle are shown in Figures 2, 3, 4 and 5, respectively. All data are expressed as means (n=10). For each response variable , an Anova was done using the Statistical Systems for Personal Computers (SAS Institute, Cary, NC, USA). The factors of interest were gender of rats (male and female) and levels of dietary isoflavones.

As expected, male rats had higher growth compared to females (345 ± 17 to 631 ± 25 vs 309 ± 15 to 365 ± 18 g/16-weeks; means $\pm$ SEM, Figure 2). In both sexes, the body weights were adversely affected at the highest two levels of dietary isoflavones (i.e. 200 mg/kg diet from soy formula and 400 mg/kg diet from Novasoy).

Levels of plasma total cholesterol in male and female rats are shown in Figure 3. Contrary to industrial claims for the hypocholesterolemic effect of soy isoflavones, dietary isoflavones had little effect on plasma total cholesterol in the rat model. The only significant ($P < 0.05$) reduction was noted in the case of male rats fed the soy formula diet which could have been partly due to the hypocholestrolemic property of soy protein in the formula diet (*30)*.

Levels of plasma isoflavones (daidzein + equol + genistein + 4-ethyl phenol) in male and female rats are shown in Figure 4. The levels of plasma isoflavones in female rats were considerably lower compared to males (0.62 ± 0.06 to 2.44 ± 0.19 vs 1.61 ± 0.15 to $9.47\pm0.67\mu$mol/L; means$\pm$SEM). There was a dose-related increase in plasma levels of isoflavones (up to five-fold). The plasma values for the soy formula diet (containing 200 mg/kg isoflavones), however, were significantly ($P < 0.05$) lower compared to that for containing 200 mg/kg diet of isoflavones from Novasoy. This suggested differences in the metabolism of isoflavones in endogenous form (soy-based infant formula) versus extracted form (Novasoy).

Effects of dietary isoflavones on the length of menstrual cycle in female rats during 6th to 8th and 12th to 14th weeks of feeding are shown in Figure 5. During 6th to 8th weeks of test, there was a significant ($P < 0.05$) steady increase in the length of estrus cycle with increasing levels of dietary isoflavones (3.73 ± 0.16 to 6.09 ± 0.44 days; means$\pm$SEM). Similarly, dietary isoflavones had a significant effect on the length of estrus cycle during 12th to 14th weeks of feeding (4.32 ± 0.18 vs 6.74 ± 0.93 to 9.59 ± 1.34 days; means$\pm$SEM). The soy-based infant formula diet providing 200 mg/kg of isoflavones, however, had no effect on the length of the estrus cycle. This observation also suggested a difference in the bioactivity of dietary isoflavones based on the source.

Based on these preliminary results, it could be concluded that the metabolism of dietary isoflavones is significantly affected by sex of rats, and that the source of dietary isoflavones (such as endogenous or extracted) may have marked effect on their potency.

Conclusions

It would be difficult for human beings to consume sufficient amounts of isoflavones from natural soy foods to reach the toxicological levels that induce pathological effects recorded in animals. However, with the recent trend towards extracting isoflavones from soy for commercialised isoflavones supplements, and because such products are not closely regulated, the potential dangerous effects from self-induced mega dosing are a concern. Moreover, there is an urgent need to address recent observations that dietary isoflavones at low concentrations may stimulate human breast cancer cells to enter the cell cycle, and can speed up division of those cells that are already cancer cells, that depend upon estrogen for their growth.

Soy-based infant formulas are considered safe and important feeding option for many infants. For over 30 years, these formulas have been fed to millions of infants in USA and Canada. Scientific data have demonstrated that infants fed soy-based formulas develop and grow normally. The safety of phytoestrogens in soy-

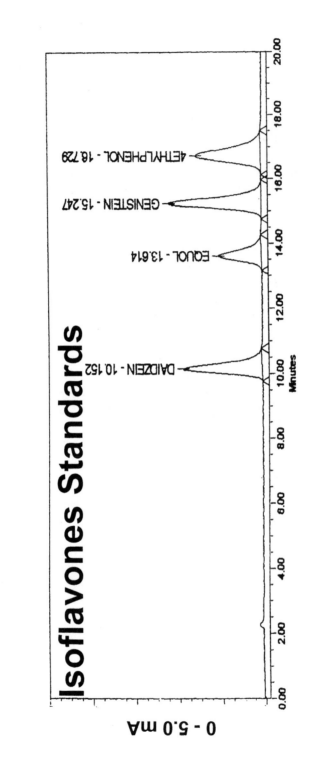

Isoflavones Standards

DAIDZEIN - 10.152
EQUOL - 13.614
GENISTEIN - 15.247
4'ETHYLPHENOL - 18.729

0 - 5.0 mA

Minutes

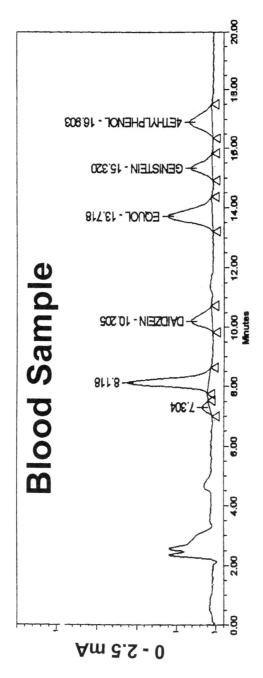

Figure 1. Liquid chromatography analysis of isoflavones in standards and in blood of rats fed a soy isoflavones-fortified diet.

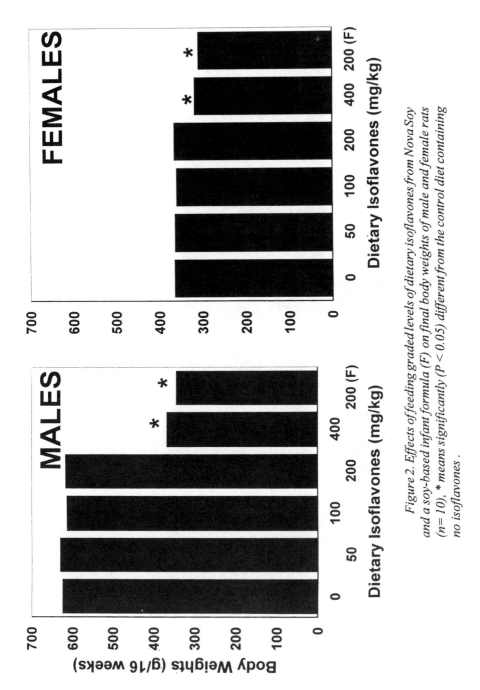

Figure 2. Effects of feeding graded levels of dietary isoflavones from Nova Soy and a soy-based infant formula (F) on final body weights of male and female rats (n = 10), * means significantly (P < 0.05) different from the control diet containing no isoflavones.

253

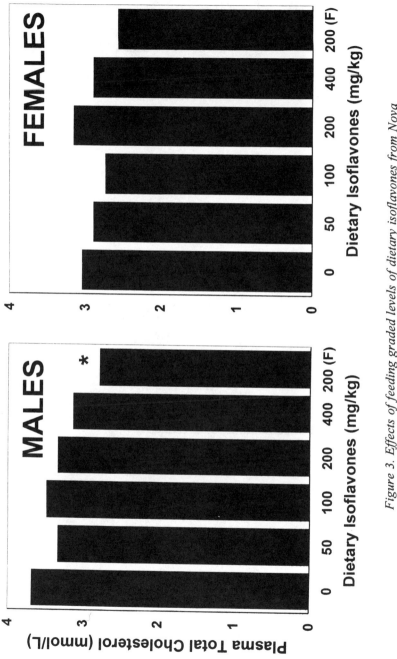

Figure 3. Effects of feeding graded levels of dietary isoflavones from Nova Soya and a soy-based infant formula (F) on the concentrations of plasma total cholesterol in male and female rats (n= 10). * means significantly (P < 0.05) different from the control diet containing no isoflavones.

254

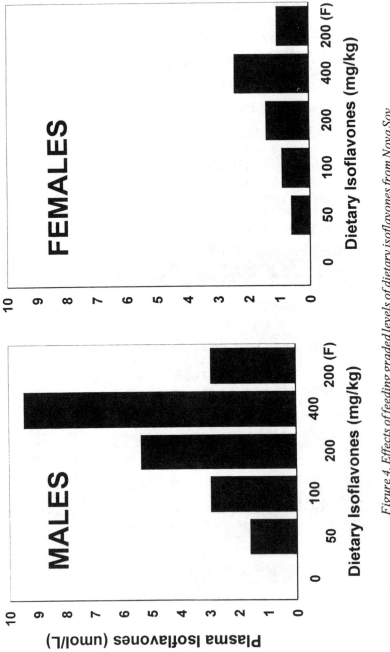

Figure 4. Effects of feeding graded levels of dietary isoflavones from Nova Soy and a soy-based infant formula (F) on the concentrations of plasma isoflavones (daidzein + equol + genistein + 4-ethyl phenol) in male and female rats (n= 10),

255

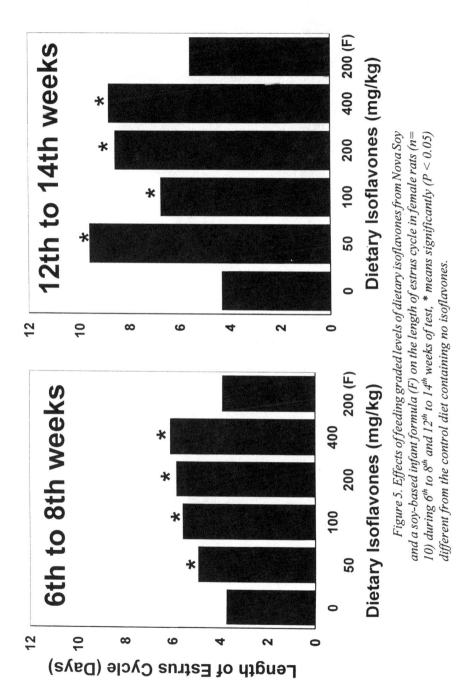

Figure 5. Effects of feeding graded levels of dietary isoflavones from Nova Soy and a soy-based infant formula (F) on the length of estrus cycle in female rats (n= 10) during 6th to 8th and 12th to 14th weeks of test, * means significantly (P < 0.05) different from the control diet containing no isoflavones.

256

based formulas has, however, been questioned because of animal-based data showing adverse endocrine effects. These clearly indicate the potential for adverse effects in humans although extrapolation of the animal data to human infants may not be considered appropriate. Nevertheless, there is a need to establish upper safe limits for isoflavones in soy-based infant formulas. In addition, it is important to further assess the safety and efficacy of dietary phytoestrogens. Information on the concentration and composition of isoflavones and their metabolites in target tissues and their metabolic effects during early development, growth and reproduction in animal models (multi-generation studies) will help provide this data.

A recent report about the association of midlife tofu consumption with brain function and structural changes later in life would suggest the need for a similar examination in adults fed soy-based formulas during infancy. Unfortunately, this information is lacking. Because of potential risk, public health agencies in England and Australia advise parents who want to use soy-based formula to first consult a doctor. In Switzerland, health authorities recommend very restrictive use of soy-based formulas for infants.

References

1. Kurzer, M.S.; Xu, X. *Annu. Rev. Nutr.* **1997**, *17*, 353.
2. Setchell, K.D.R.; Cassidy, A. *J. Nutr.* **1999**, *129*, 758S.
3. Bingham, S.A.; Atkinson, C.; Liggins, J.; Bluck, L.; Coward, A. *Br. J. Nutr.* **1998**, *79*, 393.
4. Humfrey, C.D.N. *Nat. Toxins* **1998**, 6, 51.
5. Knight, D.C.; Eden, J.A. *Obstetrics and Gynecology* **1996**, *87*, 897.
6. Tham, D.M.; Gradner, C.D.; Haskell, W.L. *J. Clin. Endocrinol. Met.* **1998**, *83*, 2223.
7. Sheehan, D.M.; Delclos, K.B.; Doerge, D.R.; Branham, W.S.; Newbold, R.R. *J. Nutr.* **2000**, *130*, 678S
8. Murkies, A.L.; Wilcox, G.; Davis, S.R. *J. Clin. Endocrinol. Met.* **1998**, *83*, 297.
9. Anderson, R.L.; Wolf, W.J. *J. Nutr.* **1995**, *125*, 581S.
10. Coward, L.; Barnes, N.C.; Setchell, K.D.R.; Barnes, S. *J. Agri. Food Chem.* **1993**, *41*, 1961.
11. Reinli, K.; Block, G. *Nutr. Cancer* **1996**, *26*, 123.
12. Sarwar, G.; L'Abbé, M.R.; Brooks, S. *J. Nutr.* **2000**, *130*, 710S.
13. Jones, A.E.; Price, K.R.; Fenwick, R.G. *J. Sci. Food Agric.* **1989**, *46*, 357.
14. Aldercreutz, H.; Honjo, H.; Higashi, A.; Fotsis, T.; Hamalainen, E.; Hasegawa, T.; Okada, T. *Am. J. Clin. Nutr.* **1991**, *54*, 1093.
15. Barnes, S.; Grubbs, C.; Setchell, K.D.R.; Carlson, J. In *Mutagens and Carciongens in the Diet;* Parzia, M., Ed.; Wiley-Liss, New York, 1990; p 239.

16. Setchell, K.D.R.; Zimmer-Nechemias, L.; Caj, J.; Heubi, J.E. *Lancet* 1997, *350*, 23.

17. Food and Drug Administration *Federal Register* 1998, 62977.

18. Food and Drug Administration *Federal Register* 1999, 57700.

19. Wu, A.H.; Ziegler, R.G.; Nomura, A.M.Y.; West, D.W.; Kolonel, L.N.; Horn-Ross, P.L.; Hoover, R.N.; Pike, M.C. *Am. J. Clin. Nutr.* 1998, *68*, 1437S.

20. Aldercreutz, H.; Mazur, W.; Bartels, P.; Elomaa, V.-V.; Watanabe, S.; Wähälä, K.; Landström, M.; Lundin, E.; Bergh, A.; Damber, J.-E.; Åman, P.; Widmark, A.; Johansson, A.; Zhang, J.-X.; Hallmans, G. *J. Nutr.* 2000, *130*, 658S.

21. Messina, M.; Erdman Jr., J.W. *J. Nutr.* 2000, *130*, 653S.

22. Bennetts, H.W.; Underwood, E.J.; Shier, F.L. *Aust. Vet. J.* 1946, *22*, 2.

23. Lindsay, D.R.; Kelly, R.W. *Aust. Vet. J.* 1970, *46*, 219.

24. Setchell, K.D.R.; Gosselin, S.J.; Welsh, M.B.; Johnson, J.O.; Balistreri, W.F.; Kramer, L.W.; Dresser, B.L.; Tarr, M.J. *Gastroent.* 1987, *93*, 225.

25. Chapin, R.E.; Stevens, J.T.; Hughes, C.L.; Kelce, W.R.; Hess, R.A.; Daston, G.P. *Fundam. Appl. Toxicol.* 1996, *29*, 1.

26. Chang, H.C.; Churchwell, M.I.; Delclos, K.B.; Newbold, R.R.; Doerge, D.R. *J. Nutr.* 2000, *130*, 1963.

27. Dees, C.; Foster, J.S.; Ahmed, S.; Wimalasena, J. *Environ. Health Perspect.* 1997, *105 (Suppl.)*, 633.

28. White, L.R.; Petrovitch, H.; Ross, G.W.; Masaki, K.; Hardman, J.; Nelson, J.; Davis, D.; Markesbery, W. *J. Am. Coll. Nutr.* 2000, *19*, 242.

29. Irvine, C.H.G.; Fitzpatrick, M.G.; Alexander, S.L. *Proc. Soc. Exp. Biol. Med.* 1998, *217*, 247.

30. Fort, P.; Moses, N.; Fasano, M.; Goldberg, T.; Lifshitz, F. *J. Am. Coll. Nutr.* 1990, *9*, 164.

31. Reeves, P.G.; Nielsen, F.H.; Fahey, G.C. *J. Nutr.* 1993, *123*, 1939.

32. Wang, H.J.; Murphy, P.A. *J. Agri. Food Chem.* 1994, *42*, 1666.

33. King, R.A.; Broadbent, J.L.; Head, R.J. *J. Nutr.* 1996, *126*, 176

34. Kieft, K.A.: Bocan, T.M.A.: Krause, B.R. *J. Lipid Res.* 1991, *32*, 859.

Chapter 18

Effect of Aqueous Extract from *Lepidium meyenii* on Mouse Behavior in Forced Swimming Test

Bo Lin Zheng[1], Kan He[1], Zhen Yen Hwang[2], Yang Lu[2],
Sui Jun Yan[3], Calvin Hyungchan Kim[1], and Qun Yi Zheng[1]

[1]Pure World Botanicals, Inc., 375 Huyler Street,
South Hackensack, NJ 07606
[2]Shenyang Medical College, Shenyang, People's republic of China
[3]Liaoning College of Traditional Chinese Medicine, Shenyang, Liaoning,
People's Republic of China

The plant *Lepidium meyenii* (Walp.), with the common name of maca, is a less known domesticated plant of Peru. In the present study, we have investigated the activity of energy enhancement of aqueous extracts from roots of maca on the behavior in mice using forced swimming test. The results of dose-response study with the aqueous extract, MacaForce™ AQ-1, in a range of daily dose at 4, 10, 20 and 40 mg/g body weight of mouse indicated that the maximum activity was reached by two weeks during a 21-day study. The energy enhancement was especially evident at the daily dose level greater than 20 mg/g body weight. The swim time in mice with 7-day oral administration with MacaForce™ AQ-1 at daily dose of 20 and 40 mg/g body weight were 13.50 ± 3.18 minutes and 16.72 ± 2.94 minutes, respectively (control group at 10.47 ± 2.38 minutes); with 14-day oral administration of MacaForce™ AQ-1 at the same doses were 17.42 ± 4.22 minutes and 19.33 ± 3.86 minutes, respectively (control group at 11.45 ± 2.46 minutes). In addition, the swim time in mice with 7-day oral administration of aqueous extracts, MacaForce™ AQ-3 and MacaForce™ AQ-4, at daily dose 40 mg/g body weight were 10.89±2.74 minutes and 14.38±3.09 minutes, respectively (control group at 11.74±1.38 minutes); with 14-day oral administration of MacaForce™ AQ-3 and AQ-4 at the same dose were 14.81±2.11 minutes and 15.34±2.60 minutes, respectively (control group at 10.90±2.04 minutes). The study with aqueous extracts, MacaForce™ AQ-1, AQ-2, AQ-3, and AQ-4, showed that the increase in swim time was directly related to the increase in content of polysaccharides in the aqueous extracts. Increase in serum lactate dehydrogenase (LDH) activity and decrease in the concentration of serum lactic acid were dose related with the content of potential active constituents.

Lepidium meyenii (Walp.), known as maca, is a plant that grows in the Andean Mountains at altitudes of more then 10,000 feet (*1*). Maca is one of the few plants that learned to flourish in this harsh environment of intense sunlight, violent winds, poor soil and low temperature. Native Peruvians have traditionally utilized maca for both nutritional and medicinal purposes. For centuries, maca roots have been used to enhance the fertility and sexual performance of men and women, as well as to treat women with menopausal symptoms. In 1960's, some studies were conducted to determine the effect of maca in rats, but the design and results were far from being satisfactory (*2*). A recent *in vivo* animal study has demonstrated that oral administration of purified lipidic extract of maca enhanced the sexual function in mice and rats (*3*). The present study used a forced swimming test followed by measurements of lactic acid, lactate dehydrogenase and melonic acids in serum to evaluate the energy enhancement property of aqueous maca extracts, MacaForce™, in mice. Unlike the lipidic extract, the major constituents of aqueous extract of maca include a number of water soluble materials, such as amino acids, trace mineral, sugars, polysaccharides, partial fatty acids and sterols. The results of the study indicated that aqueous extract of *Lepidium meyenii* (Walp.) significantly enhanced motor tolerance and relieved fatigue of the mice. The present study reveals for the first time an energy enhancement property of *Lepidium meyenii*.

Materials and Methods

Plant Material, Extraction and Purification

The dried maca roots were supplied from Peru in 1998. A voucher specimen is deposited in the Herbrio de Museo de Historia Natural 'J. Prado' Un. H. S., Lima, Peru. The extraction was carried out using ethanol and water as solvent via PureWorld's proprietary extraction and purification process (4). The resulting aqueous extracts, MacaForce™ AQ-1, AQ-2, AQ-3, and AQ-4, were formulated with excipient like maltodextrin, or with other herbs such as ginseng or rhodiola for comparison. These extracts were dried to form a powdered extract.

Phytochemical Analysis

MacaForce™ AQ-1 contains 0.004% β-sitosterol (cholestrane was used as internal standard), 0.045% fatty acids, 4.56% amino acids (0.17% alanine, 0.10% arginine, 0.03% aspartic acid, 0.27% glutamic acid, 0.05% glycine, 0.16% isoleucine, 0.14% leucine, 3.25% proline, 0.03% serine, 0.07% threonine, 0.02% tyrosine, 0.27% valine), 22.0% polysaccharide (hydrolysable carbohydrate: 1.3% glucose, 4.7% fructose and 16.0% sucrose) and 0.03% macaene and macamide.

MacaForce™ AQ-2 contains 0.18% benzyl-isothiocyanate, 0.019% sterols (0.006% campesteol, 0.003% stigmasterol, 0.010% β-sitosterol), 1.11% fatty acids (0.28% capric acid, 0.2% lauric acid, 0.19% palmitic acid, 0.02% stearic acid, 0.06% oleic acid, 0.24% linoleic acid, 0.12% linolenic acid), 5.97% amino acids (0.145% alanine, 0.374% arginine, 0.139% aspartic acid, 0.252% glutamic acid, 0.060% glycine, 0.030% histidine, 0.039% isoleucine, 0.038% leucine, 0.031% lysine, 0.013% methionine, 4.630% proline, 0.028% serine, 0.052% threonine, 0.019% tyrosine, 0.115% valine), 21.0% polysaccharide (hydrolyzable carbohydrate: 1.20% glucose, 4.45% fructose and 15.3% sucrose), and 0.27% macaene and macamide

MacaForce™ AQ-3 contains 0.001% β-sitosterol, 0.011% fatty acids, 1.137% amino acids (0.041% alanine, 0.024% arginine, 0.008% aspartic acid, 0.068% glutamic acid, 0.013% glycine, 0.040% isoleucine, 0.035% leucine, 0.811% proline, 0.008% serine, 0.018% threonine, 0.005% tyrosine, 0.066% valine), 2.853% polysaccharide (hydrolysable carbohydrate,0.157% glucose, 0.583% fructose and 2.113% sucrose), 0.006% macaene and macamide, 0.14% ginsenoside from Siberian ginseng, and 0.07% salidroside from rhodiola.

MacaForce™ AQ-4 contains 0.073% benzyl-isothiocyanate, 0.0039% sterol (0.0012% campesterol, 0.0006% stigmasterol, and 0.0021% β-sitosterol), 0.234% fatty acids (0.059% capric acid, 0.042% lauric acid, 0.040% palmitic acid, 0.004% stearic acid, 0.013% oleic acid, 0.051% linoleic acid, and 0.025% linolenic acid), 2.218% amino acid (0.056% alanine, 0.143% arginine, 0.053% aspartic acid, 0.097% glutamic acid, 0.023% glycine, 0.012% histidine, 0.015% isoleucine, 0.015% leucine, 0.012% lysine, 0.005% methionine, 1.776% proline, 0.011% serine), 9.465% polysaccharide (hydrolyzable carbohydrates, 0.544% glucose, 2.012% fructose and 6.909% sucrose), and 0.006% macaene and macamide

Animals

Male and female Shenyang mice (Grade II) weighing 25 ± 1 g were used in the study. Age of mice at the start of the experiment was 8-10 weeks. All mice had access to standard diet and water *ad libitum*. Each test article was added into 10% ethanol/water solution, and was heated to 30°C to make a suspension. The test article and a blank 10% ethanol/water solution of equal volume at daily dose specified by the experimental design were orally administered to experimental and control groups, respectively, by gavage twice per day. Each group of 15 mice consisted of about 50% each of male and female. Periodic analysis of water was performed, and there was no presence of known contaminants in the diet or water.

Forced Swim Tests

In the first study, fifteen groups of fifteen mice in each group were used. Five groups each from the fifteen groups were selected randomly for 7-day diet test, 14-day diet test, and 21-day diet test. Each five groups consisted of one control, and four groups each fed with MacaForce™ AQ-1. MacaForce™ AQ-1, an aqueous extract of *Lepidium meyenii*, was administrated to the four groups by gavage at dose of 4, 10, 20 and 40 mg/g body weight. A 4-g lead weight was attached to tail of each mouse and was placed inside a polypropylene jar (30 x 15 x 40 cm) containing 30 cm water maintained at 23 ± 2°C. The total swim time in minutes was recorded from the time the mouse was placed in water to the time of death (Table I). Additional experiments using other aqueous extracts of maca, MacaForce™ AQ-3 and MacaForce™ AQ-4, were conducted to determine their property of energy enhancement by the forced swim test. In this study, eight groups of 15 mice in each group were used. Four groups were selected randomly for 7-day diet test, and the remaining for 14-day diet test. Each four groups consisted of one control group and one group each fed with maltodextrin, MacaForce™ AQ-3, and MacaForce™ AQ-4. All test articles were administered twice daily to mice by gavage at dose of 40 mg/g body weight. On Day 7 or Day 14, at 2 hours after the second oral administration, a 4-g lead weight was attached to tail of each mouse and was placed inside a polypropylene jar (30 x 15 x 40 cm) containing 30 cm water maintained at 23 ± 2°C. The total swim time in minutes was recorded from the time the mouse was placed in water to the time of death (Table II).

Table I. Effect of *L. meyenii* Lipidic Extract MacaForce™ AQ-1 and on Swim Time of Mice in Forced Swim Test

Day	Group	Dose (per each mouse, mg/g body wt.)	Animal Number (n)	Duration of Swim Time in Minutes (mean ± SD)	P*
7	Control		15	10.47 ± 2.38	-
	MacaForce™ AQ-1	4	15	10.78 ± 2.68	>0.05
		10	15	11.14 ± 3.02	>0.05
		20	15	13.50 ± 3.18	<0.05
		40	15	16.72 ± 2.94	<0.01
14	Control		15	11.34 ± 2.46	-
	MacaForce™ AQ-1	4	15	12.63 ± 1.98	>0.05
		10	15	14.90 ± 3.35	<0.05
		20	15	17.42 ± 4.22	<0.05
		40	15	19.33 ± 3.86	<0.0
21	Control		15	10.88 ± 2.76	-
	MacaForce™ AQ-1	4	15	12.67 ± 2.94	>0.05
		10	15	15.25 ± 3.66	<0.01
		20	15	18.84 ± 3.59	<0.01
		40	15	21.37 ± 4.02	<0.0

*Vs control.

Table II. Effect of *L. meyenii* Lipidic Extract with Siberian Ginseng and Rhodiola on Swim Time of Mice in Forced Swim Test

Day	Group	Dose (per each mouse, mg/g body wt.)	Animal Number (n)	Duration of Swim Time in Minutes (mean ± SD)	P[*]
7	Control		15	11.74 ± 1.38	
	Maltodextrin	40	15	10.98 ± 2.59	>0.05
	MacaForce™ AQ-3	40	15	10.89 ± 2.74	>0.05
	MacaForce™ AQ-4	40	15	14.38 ± 3.09	<0.05
14	Control		15	10.90 ± 2.04	
	Maltodextrin	40	15	11.67 ± 2.35	>0.05
	MacaForce™ AQ-3	40	15	14.81 ± 2.11	<0.01
	MacaForce™ AQ-4	40	15	15.34 ± 2.60	<0.01

[*] vs control.

In the second study, each of MacaForce™ AQ-2 at daily dose of 40 mg/g body weight was orally administered to a group of randomly selected 15 mice by gavage twice per day for 21 days. Another group of 15 mice received a blank 10% ethanol/water solution by similar manner. On Day 21, two hours after the oral administration, a 4-g lead weight was attached to each mouse and was placed into a vertical glass cylinder (40 × 40 cm, diameter × height) containing 30 cm water maintained at 23 ± 2°C. The time at which each mouse was submerged in water for more than 3 seconds denoted the end of the swim time. Total swim time of each mouse was measured (Table III). In this study, the methodology for swim test was different compared to that of the first study in order to collect the fresh blood samples from live mice for the determination of lactic acid, LDH and MDA in blood.

Table III. Effect of *L. meyenii* lipidic Extract on Swim Time of Mice in Forced Swim Test[*]

Day	Group	Dose (per each mouse, mg/g body wt.)	Animal Number (n)	Duration of Swim Time in Seconds (mean ± SD)	P[**]
21	Control		15	110.07 ± 2.58	
	MacaForce™ AQ-2	40	15	124.07 ± 3.30	<0.01

[*]Swim time: denoted on mouse submerged in water for more than 3 seconds.
[**]vs control.

Determination of Lactic Acid in Blood

After the 30-min forced swim test on Day 21, all mice were given a 3-day rest. On Day 24, 2 hours after the second oral administration of MacaForce™ AQ-2 , each mouse free of weights was placed in 23 ± 2 water for 30-minutes. Blood samples were obtained 20 minutes before, 20 and 50 minutes after the swim test to determine the concentration of lactic acid in blood before and after the rigorous exercise (Table IV). The lactic acid analysis was performed using a spectrophotometric method with Spectrophotometer Type 201 (Shanghai Scientific, Shanghai, P.R. China) as described by FDA of China (5).

Determination of Activity of Serum Lactate Dehydrogenase (LDH)

On Day 24, blood samples were collected from mice tail before and 2 hours after the 30-min swim test. The activity of serum lactate dehydrogenase was

Table IV. Effect of *L. meyenii* LipidicExtract MacaForce™ AQ-2 on
Concentration of Serum Lactic Acid in Mice Before and After Swim Test

Group	Concentration of Lactic Acid (mmol/L)		
	20 min Before Swim Test	After 30 min Swim Test Followed by 20-min Rest	After 30 min Swim Test Followed by 50-min Rest
Before Experiment (Day 1)			
Control	5.92 ± 0.37	16.37 ± 0.78	13.07 ± 0.98
MacaForce™ AQ-2	5.87 ± 0.38	16.94 ± 0.98[a]	12.57 ± 0.38[a]
After Experiment (Day 24)			
Control	5.93 ± 0.58	16.62 ± 0.67	12.51 ± 0.84
MacaForce™ AQ-2	6.07 ± 0.83	12.13 ± 0.52[a]	6.56 ± 0.35[a]

[a] $p < 0.01$ vs control.
n = 15 for all groups.

determined by the Serum Lactate Dehydrogenase Testing Device (Beijing Zhong Sheng Biotech, Beijing, P.R. China) as described by FDA of China (5).

Determination of the Concentration of Serum Malonic Aldehyde (MDA)

On Day 24, blood samples were collected from mice tail 20 minutes before the 30-minute swim test. The concentration of MDA was determined by the barbitolthioate colorimetric method (6) using a MDA Testing Agent Device (Nanjing Jiangcheng Biotech Institute, Nanjing, P. R. China) as described by FDA of China (5).

Statistical Analysis

Pair-wise statistical comparisons between control and treated groups were done with Student's t-test. Mean differences were considered statistically significant if p < 0.05.

Results and Discussion

The results on Table I indicated that the effect of energy enhancement in mice of an aqueous extract of *Lepidium meyenii*, MacaForce™ AQ-1, evidenced by the increase in swim time, was strong. The swim time in mice with 7-day oral administration of MacaForce™ AQ-1 at daily dose of 40 mg/g body weight were 16.72±2.94 min. (Control group at 10.47±2.38 minutes). The swim time in mice with 14-day oral administration of MacaForce™ AQ-1 at the same dose were 19.33±3.86 min. (Control group at 11.45±2.46 min.). The results also indicated that the higher the dose of oral administration, the greater the effect of energy enhancement observed in mice. The energy enhancement was especially evident at the dose level greater than 20 mg/g body weight. The maximum activity was reached in 2 weeks during the 21-day study.

The results of Table II indicated that the swim time in mice with 7-day oral administration of MacaForce™ AQ-3 and MacaForce™ AQ-4 at daily dose of 40 mg/g body weight were 10.89±2.74 min. and 14.38±3.09 min., respectively (control group at 11.74±1.38 minutes). The swim time in mice with 14-day oral administration of MacaForce™ AQ-3 and AQ-4 at the same dose were 14.81±2.11 min. and 15.34±2.60 min., respectively (control group at 10.90±2.04 min.). The energy enhancement of both AQ-3 and AQ-4 were less than that of AQ-1. In addition, the results of Table II also indicated that maltodextrin used in the formulation as excipient was negative on the effect of energy enhancement.

The results of Table III indicated that the effect of energy enhancement of MacaForce™ AQ-2 (124.07±3.30 s) was stronger than the control group (110.27±2.58 s). Due to different methodology used, the results could not compare directly to the results of Tables I and II. However, in speculation of possible active components, MacaForce™ AQ-1 and MacaForce™AQ-2 with the highest polysaccharides content exhibited the greater energy enhancement effect than AQ-3 and AQ-4, which contained less polysaccharides. In short, polysaccharides may be responsible for energy enhancement effect as the major active constituents in the aqueous extracts of maca.

To further investigate the mechanism of anti-fatigue effect of maca, the concentrations of lactic acid in blood and the activity of serum lactate dehydrogenase (LDH) and the malonic aldehyde in serum were measured (Tables IV, V, VI). The two groups of mice (control and MacaForce™ AQ-2) were used. The concentration of serum lactic acid was measured on both Day 1 and Day 24 (before and after the swim test). As predicted, both the control and experimental group mice exhibited similar concentration of serum lactic acid on Day 1. On Day 24, the concentrations of serum lactic acid were significantly reduced in the experimental group mice. The concentrations of serum lactic acid after the swim test followed by a 20-minute rest were 12.13±0.52 mmol/L for the MacaForce™ AQ-2 group as compared to 16.62±0.67 mmol/L of the control group. After an additional 30-minute rest, the serum lactic acid was reduced to 6.56±0.35 mmol/L in the MacaForce™ AQ-2 group as compared to 12.51±0.84 mmol/L for the control group.

Table V. Effect of *L. meyenii* Lipidic Extract MacaForce™ AQ-2 on Change in the Activity (U/100 mL) of Serum Lactate Ddehydrogenase (LDH) in Mice 20 Minutes Before and 2 Hours After a 30-minute Swim Test

Group	Concentration of LDH (U/100mL)		
	Before Experiment	After Experiment	p^a
Control	397.4 ± 35.3	391.5 ± 56.1	
MacaForce™ AQ-2	400.8 ± 31.2	586.9 ± 42.9	< 0.01

[a]vs control.

n = 15 for all groups.

Table VI. Effect of L. meyenii Lipidic Extract MacaForce™ AQ-2 on Concentration of Serum Malonic Aldehyde (MDA) in Mice After 24-day Diet. Blood Samples Were Obtained 20 Minutes Before a 30-minute Swim Test.

Group	N	MDA (μmol/L)
Control	15	8.08 ± 0.39
MacaForce™ AQ-2	15	7.78 ± 0.43[a]

[a]$p < 0.01$ vs control.

Unlike the control group, the serum lactic acid level of the experimental mice returned to the level prior to the swim test after a 50-minute rest. The mice fed with the purified maca aqueous extracts, MacaForce™ AQ-2 for 24 days were more successful in faster reduction of the serum lactic acid level. This is evidenced by higher level of LDH activity found in the MacaForce™ AQ-2. The LDH activities of the experimental groups, after the 24-day diet with MacaForce™ AQ-2 were increased to 586.9 ± 42.9 U/100 mL in the MacaForce™AQ-2 group as compared to 391.5 ± 56.1 U/100 mL in the control group. This corresponds to 1.5-fold increase in the LDH activity in mice treated with MacaForce™ AQ-2. The increase in activity of LDH was the reason for the reduction of the concentration of lactic acid in blood. The high content of polysaccharides or other constituent may be responsible for the higher LDH activity observed. The present study showed that the increase in LDH activity and the decrease in concentration of serum lactic acid were dose related with the content of the polysaccharides.

The concentration of serum malonic aldehyde (MDA) was measured here as an indication of the anti-fatigues as well as anti-aging effect of the Maca extract. MDA is a final product of cellular peroxidation. The concentration of serum

268

MDA corresponds to the quantity of free radicals, which causes aging, fatigue, and even death (7). The purified maca lipidic extracts significantly reduced serum MDA level from 8.08±0.39 μmol/L in mice of control group to 7.78±0.43 μmol/L in mice of MacaForce™ AQ-2 group.

During a strenuous physical exercise, the muscular glycogen of mice was hydrolyzed to lactic acid by enzyme, thereby increasing the level of lactic acid concentration and decreasing the pH of blood. The change in pH decreased the binding affinity of muscular proteins to Ca^{2+} and inhibited the contraction of the muscles, thereby causing fatigue (6). The results from this *in vivo* study showed that the MacaForce™ AQ products exhibited significant anti-fatigue effect in the mice. From a perspective of phytochemistry, the experimental data supported the bioactivities of the extracts in the dose response form to the content of the polysaccharides. The higher the content of the polysaccharide the stronger anti-fatigue effect was observed. Further studies to identify the major active constituents responsible for the anti-fatigue effect in mice as well as their mechanism of action are in progress. The present study reveals for the first time an anti-fatigue effect of maca.

References

1. Leon, T. *Economic Botany* **1964**, *18*, 122-127.
2. Chacón, R.C. Thesis, University Nac. Mayor de San Marcos, Lima, Peru, 1961, p. 43.
3. Zheng, B. L.; He, K.; Kim, C. H.; Rogers, L. L.; Yu, S.; Huang, Z. Y.; Lu, Y.; Yan, S. J.; Qien, L. C.; Zheng, Q. Y *Urology* **2000**, *55*, 578-602.
4. Zheng, B. L.; Kim, C. H.; Wolthoff, S.; He, K.; Shao, Y.; Rogers, L. L.; Zheng, Q. Y. Patent pending, 1999.
5. FDA of China, *National Clinical Test Procedure*, 1998.
6. Chi, F. J. *The Journal of 4th Military Medical University* **1986**, *6*, 152.
7. Xu, S. K. *Antiaging*, Chinese Medicinal and Pharmaceutical Press, Beijing, 1994, p. 640.

Chapter 19

Bioactive Homoisoflavones from Vietnamese Coriander or Pak Pai (*Polygonatum odoratum*)

Bret C. Vastano[1,2], M. Mohamed Rafi[3], Robert S. DiPaola[3],
Nanqun Zhu[2], Chi-Tang Ho[1,2], Anthony T. Rella[1,2],
Geetha Ghai[1], and Robert T. Rosen[1,*]

[1]Center for Advanced Food Technology, and [2]Department of Food Science,
Cook College, Rutgers, The State University of New Jersey,
63 Dudley Road, New Brunswick, NJ 08901–8520
[3]The Cancer Institute of New Jersey, University of Medicine and Dentistry
of New Jersey, Little Albany Street, New Brunswick, NJ 08901

The roots of *Polygonatum odoratum* were screened for
compounds which induce apoptosis in breast cancer cell lines.
The powdered roots were extracted with methanol and ethyl
acetate. The ethyl acetate fraction was then subjected to
bioassay directed fractionation using silica gel column
chromatography. Bcl-2 protein, a regulator of apoptosis, was
assessed by immunoblot. The active fraction was eluted with
30:1 $CHCl_3/CH_3OH$ and was rechromatographed on a second
silica gel column. Elution with 20:80 hexanes/chloroform
yielded the active fraction which was purified using
semipreparative HPLC. Two compounds were identified.
These include 2,3-dihydro-3-[(15-hydroxyphenyl)methyl]-
5,7-dihydroxy-6-methyl-8-methoxy-4H-1-benzopyran-4-one
and 2,3-dihydro-3-[(15-hydroxyphenyl)-methyl]-5,7-di-
hydroxy-6,8-dimethyl-4H-1-benzopyran-4-one.

The roots of *Polygonatum odoratum* have been used in traditional Chinese medicine for a variety of therapeutic purposes (*1*). It has been used as a crude medicinal agent in the treatment of analeptic (*2*) and as a nutritious tonic in Asia (*3*). In addition to China, *P. odoratum* grows in Thailand and Vietnam where it is known as Pak pai or Vietnamese mint. It can also be found growing throughout the southern United States. *P. odoratum* is commonly used as a condiment. Its methanol extract has been shown to suppress 96% of the mutagenicity of Trp-P-1 (*4*). *P. odoratum* alcohol extracts have also served as an immunopotentiator of mice injured by burns (*5*). Compounds that have been previously identified in *P. odoratum* include steroidal saponins (*1*), other steroidal compounds (*3*), quercitol (*6*), azetidine 2-carboxylic acid (*7-8*), mucous polysaccharides (*2*), vitamin A, mucilage (*9*) and diosgenin (*10*).

The bcl-2 protein (Molecular Weight about 26 kDa) is a member of cytoplasmic proteins which regulate cell death (*11*). Bcl-2 is known to be over-expressed in tumor cells and has been shown to promote cell survival (*12*) by inhibiting the process of cell death, also known as apoptosis (*13,14*). Where bcl-2 acts to inhibit apoptosis, bax, a second member of cytoplasmic proteins, counteracts this protective effect (*15*). Hunter proposes that bcl-2 protects cells from apoptosis by dimerizing with bax (*15*). The phosporylation of bcl-2 (*16*) interferes with its dimerization to bax resulting in more bax homodimers and subsequent apoptosis (*15*). Current active chemotherapy agents, such as taxol, vincristine, and vinblastine can induce bcl-2 phosphorylation in cancer cells and cause programmed cell death (*15,17*). Since bcl-2 expression in tumor cells may be responsible for chemotherapy resistance, novel agents that abrogate bcl-2 function by inducing phosphorylation are attractive candidates for study.

This research involves bioassay-directed fractionation of *Polygonatum odoratum*, whose crude extracts were found to phosphorylate bcl-2 protein in breast cancer cell lines.

Materials and Methods

General Procedures

^{1}H NMR and ^{13}C NMR spectra were obtained on a VXR-200 instrument and mass spectra were obtained using atmospheric pressure chemical ionization (APCI) in the negative-ion mode and by direct probe electron ionization (EI). APCI MS analysis was performed on a Micromass Platform II system

(Micromass Co., Beverly, MA) equipped with a Digital DECPc XL560 computer for analysis of data. The ion source temperature was set at 150 °C and the probe temperature was set at 450 °C. The sample cone voltage was 10 V and the corona discharge was 3.2 kV. Direct probe EI-MS was performed on a Finnigan MAT 8230 high resolution mass spectrometer. HPLC analysis was performed on a Varian Vista 5500 Liquid Chromatograph pump coupled to a Varian 9065 Polychrom diode array detector (Sugar Land, TX). Fractionation of purified compounds was obtained on a Waters 600E HPLC pump (Milford, MA) coupled to a Milton Roy Spectro Monitor 3100 variable wavelength detector (Riviera Beach, Florida). Column chromatography was performed using a glass chromatography column purchased from Kontes (Vineland, NJ). Selecto Scientific silica gel (100-200 mesh particle size) was used for column chromatography. All fractions were screened on Whatman silica gel thin-layer chromatography (TLC) plates (250 μm thickness, 60 Å silica gel medium) with compounds revealed under fluorescent light. The column packing and TLC plates were both purchased from Fisher Scientific (Springfield, NJ). All solvents used for extraction and isolation were of HPLC grade and purchased from Fisher Scientific.

Plant Material and Cell Line

The roots of *Polygonatum odoratum* were imported from Guangdong, China and purchased from Jasmine Enterprises located in Flagstaff, AZ. MCF-7 breast tumor cells were obtained from the American Type Culture Collection (ATCC). Cells were maintained at 37°C in an atmosphere of 5% CO_2 and grown in Roswell Park Memorial Institute (RPMI) with 10% phosphate buffered solution (PBS), penicillin and streptomycin. Cells were routinely checked and found to be free of contamination by mycoplasma.

Extraction and Isolation Procedures

The roots were dried prior to being ground into a powder. The powdered roots of *P. odoratum* were extracted with methanol and concentrated under vacuum using rotary evaporation. The remaining concentrate was then partitioned with acidified ethyl acetate (3% HCl). The ethyl acetate extract of *P. odoratum* was determined to have strong bioactivity due to its ability to induce phosphorylation of the anti-apoptotic protein bcl-2 with subsequent cell death (apoptosis) in breast cancer cell lines.

The dry ethyl acetate extract was then chromatographed on a silica gel column (2.5 x 30 cm) such that bio-assay directed fractionation could be performed. The column was packed in methanol and then conditioned with 30:1 chloroform/methanol (500 mL). Elution was done using a solvent mixture of chloroform/methanol with an increasing amount of methanol (30:1, 20:1, 10:1, 8:1, 7:1, 5:1, 3:1, 1:1, 1:5, 1:15, 1:25, 0:1; each 500 mL) (*18*). Successive fractions were collected and tested for biological activity. The fraction eluted with 30:1 chloroform/methanol was determined to be most active in that it phosphorylated bcl-2 and induced apoptosis in breast cancer cell lines. This fraction was then rechromatographed on a silica gel column (2.5 x 30 cm). The column was packed in hexanes and then conditioned with 50:50 hexanes/chloroform (500 mL). Elution was carried out using a solvent mixture of hexanes/chloroform/methanol with increasing amounts of chloroform and methanol (50:50:0, 40:60:0, 30:70:0, 20:80:0, 0:100:0, 0:90:10, 0:0:100; each 500 mL). Successive fractions were collected and sent for additional biological testing.

It was determined that the fraction eluted with 20:80:0 hexanes/ chloroform/methanol was the active fraction in that it was capable of phosphorylating bcl-2 and inducing apoptosis in breast cancer cell lines. This fraction was then screened for purity using silica gel TLC plates and analytical HPLC. Separation was performed on a Discovery C18 reversed phase column (250 mm x 4.6 mm, 5 μm) with a column guard purchased from Supelco (Bellefonte, PA). The solvent program was a gradient system: A, water with 0.15% triethylamine (TEA) purchased from Sigma Chemical Co. and 0.18% formic acid (FA) purchased from Fisher Scientific; B, acetonitrile. The elution program at 1 mL min^{-1} was as follows: 100% A to 100% B (0-35 min); 100% B (35-55 min). The wavelengths monitored were 220-320 nm with a Varian 9065 diode array detector. Final separation of pure compounds was obtained using semi-preparative HPLC on a Zorbax Rx-C18 reversed phase column (9.4 mm x 240 mm, 5 μm) purchased from Mac-Mod Analytical (Chadds Ford, PA). Compounds were eluted by an isocratic solvent system containing 60% water with 0.15% TEA and 0.18% FA; 40% acetonitrile. The flow rate was at 4 mL min^{-1} and the wavelength monitored was 254 nm.

Western Blot Analysis.

Analysis of bcl-2 protein phosphorylation was performed by immunoblot as previously described (*15*). Cells were initially treated with various fractions and pure compounds. Cells were then lysed and equivalent amounts of proteins were electrophoresed by 12% sodium dodecyl sulfate-polyacrylamide gel electrophoresis (SDS-PAGE) and transferred to nitrocellulose. Bcl-2 protein was detected using a monoclonal bcl-2 primary antibody and a secondary goat anti-mouse horseradish peroxidase conjugated antibody followed by enhanced chemiluminescence (ECL) detection (*15*).

Results and Discussion

The ethyl acetate extract of *Polygonatum odoratum* was subjected to bioassay directed fractionation and yielded an active fraction upon elution with 30:1 chloroform/methanol. This fraction was then rechromatographed on a second silica gel column and elution with 20:80:0 hexanes/ chloroform/methanol yielded the active fraction. This fraction was evaporated to dryness under nitrogen at room temperature. The sample was then reconstituted in methanol and analyzed by reversed phase HPLC. The HPLC conditions are shown in the Materials and Methods section. The HPLC chromatogram of the 20:80:0 hexanes/chloroform/methanol fraction at 302 nm is presented in Figure 1.

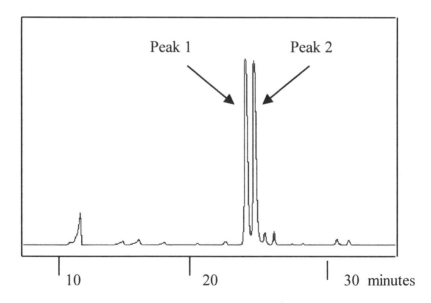

Figure 1. Reversed Phase HPLC chromatogram of Polygonum odoratum 20:80 hexane/chloroform fraction at 302 nm.

274

The 20:80:0 hexanes/chloroform/methanol fraction was also analyzed by APCI LC-MS in the negative ion mode. It was determined that peak 1 had a molecular weight of 330 as evident by the pseudomolecular ion at m/z 329 ([M-H]⁻), as shown in Figure 2. Peak 2 had a molecular weight of 314 and gave a pseudomolecular ion at m/z 313 ([M-H]⁻), as shown in Figure 2. Both compounds were tentatively identified as dihydrobenzopyranones.

For structure determination, isolation and purification of the 20:80:0 hexanes/chloroform/methanol fraction was necessary for NMR studies. The final purification of the dihydrobenzopyranones was performed using reversed phase, semi-preparative HPLC. The HPLC parameters for purification are shown in the Materials and Methods section. Peak 1 was determined to be 2,3-dihydro-3-[(15-hydroxyphenyl)methyl]-5,7-dihydroxy-6-methyl-8-methoxy-4H-1-benzopyran-4-one and Peak 2 was determined to be 2,3-dihydro-3-[(15-hydroxyphenyl)methyl]-5,7-dihydroxy-6,8-dimethyl-4H-1-benzopyran-4-one.

The structures of both dihydrobenzopyranones, as shown in Figure 3, were determined by NMR studies and EI-MS. The fragmentation patterns of both Peak 1 and Peak 2 were determined by EI-MS, shown in Figure 4. The concentration of both dihydrobenzopyranones in the powdered roots of *P. odoratum* was approximately 30 µg/g.

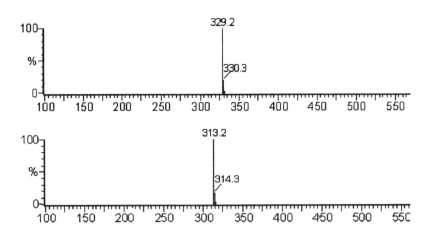

Figure 2. *APCI negative ion mass spectrum of Peak 1, m/z 329 = [M-H] and Peak 2, m/z 313 = [M-H]*

Structure Determination of Isolated Compounds

Peak 1

2,3-Dihydro-3-[(15-hydroxyphenyl)methyl]-5,7-dihydroxy-6-methyl-8-methoxy-4H-1-benzopyran-4-one: High Res. MS Experimental = 330.110648 and Theoretical = 330.110339, $C_{18}H_{18}O_6$; EI-MS: m/z = 330 [M$\overset{+}{\cdot}$], 224 ([M$\overset{+}{\cdot}$ − substituted tropylium ion + H]), 209 (m/z 224 − CH$_3$), and 107 (base peak) (substituted tropylium ion); APCI, m/z 329 ([M-H]$^-$);

^{1}H NMR δ 1.97 (3H, s, 18-H), δ 2.68 (1H, dd, J = 13.5, 10.3 Hz, 11a-H), δ 2.81 (1H, m, 3-H), δ 3.10 (1H, dd, J = 13.6, 4.2 Hz, 11b-H), δ 3.73 (3H, s, 19-H), δ 4.15 (1H, dd, J = 11.1, 6.7 Hz, 2a-H), δ 4.30 (1H, dd, J = 11.1, 3.9 Hz, 2b-H), δ 6.74 (2H, d, J = 8.2 Hz, 14, 15-H), δ 7.07 (2H, d, J = 8.2 Hz, 13, 17-H), ^{13}C NMR δ 7.52 (C-18), δ 33.23 (C-11), δ 47.80 (C-3), δ 61.84 (C-19), δ 70.39 (C-2), δ 102.34 (C-10), δ 105.28 (C-6), δ 116.40 (C-14, 16), δ 129.16 (C-8), δ 130.15 (C-12), δ 131.16 (C-13, 17), δ 152.95 (C-9), δ 157.21 (C-15), δ 158.63 (C-7)*, δ 159.20 (C-5)*, δ 199.63 (C-4). (*Assignments may be interchanged with one another)

This compound was deduced as having an elemental formula of $C_{18}H_{18}O_6$ by the ^{13}C NMR spectrum, the APCI LC-MS, which exhibited a pseudo-molecular ion at m/z 329 ([M-H]$^-$), and the high resolution EI-MS. The ^{1}H NMR spectrum showed two doublets at δ 6.74 (2H, d, J = 8.2 Hz) and δ 7.07 (2H, d, J = 8.2 Hz), due to the protons of a *p*-disubstituted B ring. The ^{13}C NMR spectrum also showed the corresponding carbon signals ($δ_C$ 157.21, 131.6, 130.15, and 116.40). In addition, the ^{13}C NMR spectra gave six carbon signals assigned for a hexasubstituted benzoyl group (A ring) ($δ_C$ 159.20, 158.63, 152.95, 129.16, 105.28, and 102.34). This result corresponded to the absence of any other proton signals but those for the B ring. Besides the signals above, the ^{1}H and ^{13}C NMR spectra exhibited the presence of one methyl group substituted on the A ring ($δ_H$ 1.97, s; $δ_C$ 7.52), one methoxy group ($δ_H$ 3.73, S; $δ_C$ 61.84), one carbonyl group ($δ_C$ 199.63), one oxygenated methylene group ($δ_H$ 4.15, 4.30, 1 H each, dd; $δ_C$ 70.39), one methylene group ($δ_H$ 2.68, 3.10, 1 H each, dd; $δ_C$ 33.23), and one methine group ($δ_H$ 2.81, m, $δ_C$ 47.80).

According to the information above, this compound has an unusual skeleton. The ^{1}H-^{1}H COSY and HMBC NMR spectra showed the presence of two moieties: (A ring)-O-CH$_2$(2)-CH(3)-CH$_2$(11)-(B ring) and (A ring)-CO(4)-CH(3)-. These moieties were connected by the following long range ^{1}H-^{13}C correlation ions: C-3/H-11, C-4/H-2, C-11/H-2, and C-12/H-11. More over, since the chemical shift of C-4 was 199.63, one hydroxyl group should be substituted at C-5. Finally, the substitution sites of the methoxyl group and the

276

methyl group were determined by the information from the ^{13}C NMR and HMBC spectra. These spectra indicated that the hydrogens from the methoxy group at H-19 were correlated with C-8 and the hydrogens from the methyl group at H-18 were correlated with C-6.

Peak 1

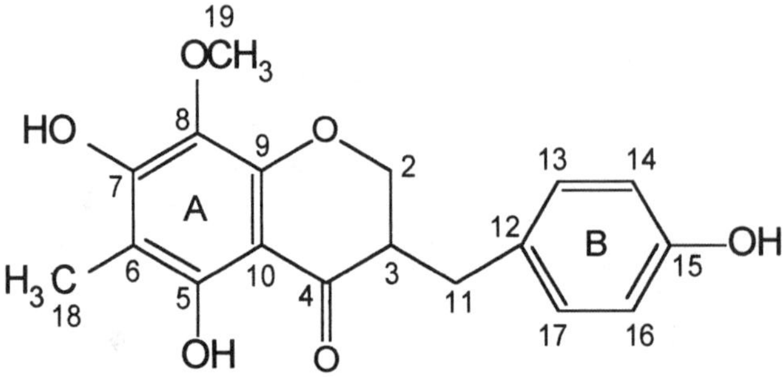

2,3-dihydro-3-[(15-hydroxyphenyl)methyl]-5,7-dihydroxy-6-methyl-8-methoxy-4H-1-benzopyran-4-one

Peak 2

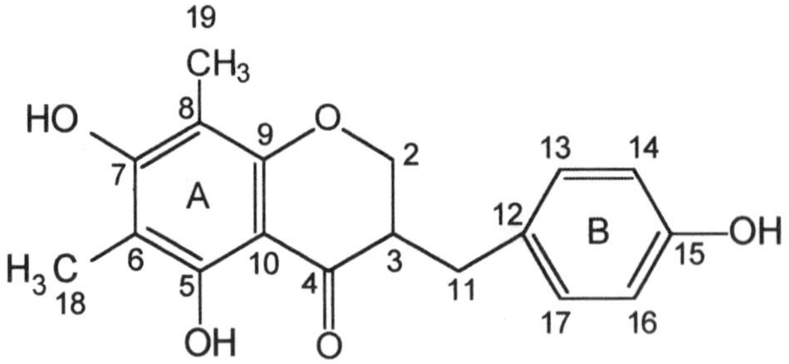

2,3-dihydro-3-[(15-hydroxyphenyl)methyl]-5,7-dihydroxy-6,8-dimethyl-4H-1-benzopyran-4-one

Figure 3. Structures of dihydrobenzopyranones isolated from Polygonum odoratum

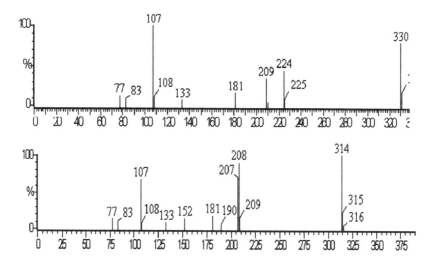

Figure 4. EI mass spectrum of Peak 1, m/z = 330 [M⁺˙], 224 ([M⁺˙ − substituted tropylium ion + H]), 209 (m/z 224 − CH₃), and 107 (base peak) (substituted tropylium ion) and peak 2, m/z = 314 [M⁺˙] (base peak), 208

Peak 2

2,3-Dihydro-3-[(15-hydroxyphenyl)methyl]-5,7-dihydroxy-6,8-dimethyl-4H-1-benzopyran-4-one: High Res. MS Experimental = 314.115729 and Theoretical = 314.115424, $C_{18}H_{18}O_5$; EI-MS: m/z = 314 [M⁺˙] (base peak), 208 ([M⁺˙ − substituted tropylium ion]), and 107 (substituted tropylium ion); APCI, m/z 313 ([M-H]⁻);

¹H NMR δ 1.98 (3H, S, 18-H)*, δ 2.00 (3H, s, 19-H)*, δ 2.69 (1H, dd, J = 13.6, 10.2 Hz, 11a-H), δ 2.83 (1H, m, 3-H), δ 3.09 (1H, dd, J = 13.6, 4.0 Hz, 11b-H), δ 4.12 (1H, dd, J = 11.5, 6.8 Hz, 2a-H), δ 4.26 (1H, dd, J = 11.5, 4.0 Hz, 2b-H), δ 6.74 (2H, d, J = 8.4 Hz, 14, 15-H), δ 7.07 (2H, d, J = 8.4 Hz, 13, 17-H), ¹³C NMR δ 7.71 (C-18)#, δ 8.17 (C-19)#, δ 33.45 (C-11), δ 48.00 (C-3), δ 70.33 (C-2), δ 102.04 (C-10), δ 104.24 (C-6)$, δ 106.36 (C-8)$, δ 116.66 (C-14, 16), δ 130.61 (C-12), δ 131.43 (C-13, 17), δ 157.64 (C-15), δ 164.25 (C-9), δ 168.20 (C-5)ᐃ, δ 168.27 (C-7)ᐃ, δ 199.60 (C-4).

(*ᐃ$# Correlating assignments may be interchanged with one another.)

The molecular weight for this compound corresponds to $C_{18}H_{18}O_5$, which was confirmed by ^{13}C NMR and by the APCI LC-MS spectrum, which gave a pseudomolecular ion at m/z 313 ([M-H]$^-$). A comparison of the ^1H and ^{13}C NMR spectra of this compound to those of Peak 1 differ only by the absence of the methoxyl group and the presence of an additional methyl group (δ_H 2.00 or 1.98; δ_C 7.71 or 8.17) on the A ring. The corresponding changes also occurred: the chemical shifts of C-5, C-7, and C-9 shifted from 159.20, 158.63, and 152.95 in Peak 1 to 168.20, 168.27, and 164.25 in Peak 2. This data is in agreement with previous reports of this compound (*19*).

Conclusion

Extensive bioassay directed fractionation has led to the isolation and identification of two dihydrobenzopyranones. Dihydrobenzopyranones, also known as homoisoflavones or 3-benzyl-4-chromanones, are a small class of phenolic compounds which have been isolated from various genera including those of the Lilaceae family (*20*). Several dihydrobenzopyranone analogues have been identified in the bulbs of *Muscari comosum* (*21-22*), *Eucomis bicolor* (*23*), *Eucomis autumnalis* (*24*) and *Eucomis punctata* (*25*), as well as in the heartwood of *Pterocarpus marsupium* (*26*). These compounds are known to be antimutagenic (*27*), antiinflammatory (*28*), and of chemotaxonomic interest (*20*).

It is of interest to note the observation of a methylene group between the A and B rings. This differs form an isoflavone where the A and B rings are directly connected. We postulate that the addition of the methylene group changes the conformation of the molecule and positively effects bioactivity when compared to the isoflavone genistein, which did not have activity in this assay. Additional moieties substituting for the methylene may also have increased bioactivity. Bioassays of the purified compounds determined that Peak 2 had increased cytotoxicity and induced bcl-2 phosphorylation, where as Peak 1 had less cytotoxicity and did not induce bcl-2 phosphorylation. Hopefully these analogs will serve as lead compounds in the synthesis of more potent analogs. This could ultimately lead to the development of improved therapeutic agents for cancer treatment, which have specific molecular targets.

Acknowledgments

CAFT is an initiative of The NJ Commission of Science and Research. This is New Jersey Agricultural Experiment Station publication # D99101-01-01. The work was supported by the New Jersey Commission on Cancer Research, NCI CA 80654, Cap Cure, and CA 77135 and the New Jersey Commission of Science and Research.

References

1. Lin, H.W.; Han, G.Y.; Liao, S.X. *Yao Hsueh Hsueh Pao*. **1994**, *29*, 215-222.
2. Tomoda, M.; Yoshiko, Y.; Tanaka, H.; Uno, M. *Chem. Pharm. Bull.* **1971**, *19*, 2173-2177.
3. Sugiyama, M.; Nakano, K.; Tomimatsu, T.; Nohara, T. *Chem. Pharm. Bull.* **1984**, *32*, 1365-1372.
4. Japan International Research Center for Agricultural Sciences (JIRCAS). **1998**, *17*, 1-3.
5. Xiao, J.; Cui, F.; Ning, T.; Zhao, W. *Chung Kuo Chung Yau Tsa Chih.* **1990**, *15*, 557-578.
6. Lazer, M.; Gheta, D.; Grigorescu, E. *Farmacia.* **1971**, *19*, 31-38.
7. Virtanen, A.I.; Linko, P. *Acta. Chem. Scand.* **1955**, *9*, 551-554.
8. Fowden, L. *Nature.* **1955**, *176*, 347-348.
9. Gaal, B. *Ungar. Pharm. Ges.* **1927**, *3*, 133-139.
10. Okanishi, T.; Akahori, A.; Yasuda, F.; Takeuchi, Y.; Iwao, T. *Chem. Pharm. Bull.* **1975**, *23*, 575-579.
11. Hunter, J.J.; Parslow, T.G. *J. Biol. Chem.* **1996**, *271*, 8521-8524.
12. Vaux, D.L.; Cory, S.; Adams, J.M. *Nature (Lond.).* **1988**, *335*, 440-442.
13. Haldar, S.; Jena, N.; DuBois, G.; Takayama, S.; Reed, J.C.; Fu, S.S.; Croce, C.M. *Arch. Biochem. Biophys.* **1994**, *315*, 483-488.
14. Miyashita, T.; Reed, J.C. *Blood.* **1993**, *11*, 151-157.
15. Haldar, S.; Chintapalli, J.; Croce, C.M. *Cancer Res.* **1996**, *56*, 1253-1255.
16. Haldar, S.; Jena, N.; Croce, C.M. *Proc. Natl. Acad. Sci.* **1995**, *92*, 4507-4511.
17. Haldar, S.; Basu, A.; Croce, C.M. *Cancer Res.* **1997**, *57*, 229-233.
18. Chen, Y.; Wang, M.; Rosen, R.T.; Ho, C.-T. *J. Agric. Food Chem.* **1999**, *47*, 2226-2228.
19. Huang, P.L.; Gan, K.H.; Wu, R.R.; Lin, C.N. *Phytochemistry* **1997**, *44*, 1369-73.
20. Heller W, Tamm C. *Fortschr. Chem. Org. Naturst.* **1981**, *40*, 105-152.
21. Adinolfi, M.; Barone, G.; Belardini, M.; Lanzetta, R.; Laonigro, G.; Parrilli, M. *Phytochemistry.* **1984**, *9*, 2091-2093.
22. Adinolfi, M.; Barone, G.; Lanzetta, R.; Laonigro, G.; Mangoni, L.; Parrilli, M. *Phytochemistry.* **1985**, *24*, 624-626.
23. Bohler, P.; Tamm, CH. *Tetrahedron Letters.* **1967**, 3479-3483.
24. Sidwell, W.T.L.; Tamm, CH. *Tetrahedron Letters.* **1970**, 475-478.
25. Finckh, R.E.; Tamm, CH. *Experientia.* **1970**, *26*, 472-473.

26. Jain, S.C.; Sharma, S.K.; Kumar, R.; Rajwanshi, V.K.; Babu, R. *Phytochemistry* **1997**, *44*, 765-766.
27. Wall, M.E.; Wami, M.C.; Manikumar, G.; Taylor, H.; McGivney, R. *J. Nat. Prod.* **1989**, *52*, 774-778.
28. Della, L.R.; Del, N.P.; Tubaro, A.; Barone, G.; Parrilli, M. *Planta Medica* **1989**, *55*, 587-588.

Chapter 20

Analysis of Bioactive Ferulates from Gum Guggul *(Commiphora wightii)*

Nanqun Zhu[1], M. Mohamed Rafi[2], Dajie Li[3], Edmond J. LaVoie[3], Robert S. DiPaola[2], Vladimir Badmaev[4], Geetha Ghai[1], Robert T. Rosen[1], and Chi-Tang Ho[1]

[1]Department of Food Science and Center for Advanced Food Technology, Rutgers University, 65 Dudley Road, New Brunswick, NJ 08901
[2]Cancer Institute of New Jersey, University of Medicine and Dentistry of New Jersey, Little Albany Street, New Brunswick, NJ 08901
[3]Department of Pharmaceutical Chemistry, College of Pharmacy, Rutgers University, Piscataway, NJ 08854
[4]Sabinsa Corp., 121 Ethel Road, Piscataway, NJ 08854

For centuries, guggul has been used extensively by Ayurvedic physicians to treat a variety of diseases, especially for disorders of lipid metabolism. In the present study, major constituents in freshly prepared guggul gum were isolated and identified. Furthermore, bioactivity-directed fractionation and purification afforded three cytotoxic components (D1, D2, and D3) of fresh *Commiphora wightii*. D1 was characterized as a mixture of two novel ferulates with an unusual skeleton by spectral and chemical methods, including NMR, GC-MS and chemical derivatization. Similarly, D2, as well as D3, was identified as the novel ferulate with the same skeleton as D1. D1 exhibited significant cytotoxic activity in detailed activity test, and showed moderate scavenging effect against 2,2-diphenyl-1-picrylhydrazyl (DPPH) radicals.

Introduction

Commiphora wightii (Burseraceae) is a branched shrub or a small tree (2 – 3 m high) found in some states of India and Pakistan. As the golden brown or reddish brown oleogum exudates of *Commiphora wightii*, guggul gum is mentioned in the classic Ayruvedic literature as an efficacious treatment for bone fractures, arthritis, inflammation, obesity, cardiovascular disease, and lipid disorders. Guggul lipid, a mixture of lipid steroids isolated from the resin of *C. wightii* has been available in the Indian market since 1988 as a potent hypolipidemic agent. Another unique feature of guggul is that major Ayurvedic treatises very emphatically stress that most of the therapeutic properties relating to fat disorders attributed to 'aged' samples of guggul. The fresh guggul gum is said to have the opposite effects (*1*).

With the discovery of the hypolipidemic activity of the gum resin, some chemical investigations have been reported. It was found that guggul resin is a complex mixture of various classes of chemical compounds, such as lignans, lipids, diterpenoids and steroids. Especially, a waxy solid, which is a mixture of esters based on homologous long chain tetraols and ferulic acid with a unique structure, was reported (*1*). Through a more detailed study by Kumar and Dev (*2*), the absolute stereochemistry of these tetraols was determined.

Few chemical studies have been done on fresh guggul gum. Here, we report 10 known compounds isolated from freshly prepared guggul gum, all of them have been reported previously in guggul gum. As part of a search for bioactive substances from natural products, it was found that the EtOAc extract of *C. wightii* showed significant cytotoxicity *in vitro* studies. Thus, directed by cytotoxic activity screening, active constituents of this plant were isolated and identified.

Materials and Methods

Material

TLC was performed on Sigma-Aldrich silica gel TLC plates (250 μm thickness, 2-25 μm particle size), with compounds visualized by UV$_{365nm}$ light and spraying with 10% (v/v) H_2SO_4 ethanol solution. Silica gel (130-270 mesh), Sephadex LH-20 and RP-18 (60 μm) (Sigma Chemical Co., St. Louis,

MO) were used for column chromatography. Potassium hydroxide, bis-(trimethylsilyl)-trifluoroacetamide, trimethylchlorosilane, dimethyl disulfide (DMDS), iodine, and sodium bisulfite were purchased from Sigma Chemical Co. (St. Louis, MO). All solvents used were purchased from Fisher Scientific (Springfield, NJ).

Instrumentation

^{1}H NMR and ^{13}C NMR spectra were obtained on a VXR-200 instrument and recorded in CDCl$_3$ with TMS as internal reference. Signals are reported in ppm (δ). 2D NMR spectra were recorded on a U-500 instrument (Varian Inc., Melbourne, Australia). Methanol-d_4 was used as solvent, and chemical shifts are expressed in parts per million (δ) using TMS as internal standard. GC and GC-MS analysis were performed on a HP 6890 GC, which is equipped with a HP 5973 Mass Selective Detector. Mass spectra were obtained in the electronic ionization (EI) mode at 70 eV. The oven temperature was 100°C, raised to 200°C at a rate of 2°C / min, held for 30 minutes, and finally raised to 280°C at 10°C / min, the injector temperature was 270°C and the detector temperature was 280°C. The split ratio of sample was set at 60:1. LC-MS data were obtained on a HP 59980 B particle beam LC/MS interface with HP 1090 HPLC instrument.

Plant Material

The exudate of *C. wightii* was obtained from Sabinsa Co. (Piscataway, NJ) and was imported from India.

Extraction and Purification

Guggul gum (2.5 Kg) was extracted with ethyl acetate at room temperature for one week. The residue was filtered and the ethyl acetate extracts were combined and concentrated under reduced pressure (850 g). A sample (100 g) was directly subjected to cc on Si gel, eluted with a solvent mixture of CHCl$_3$/MeOH with increasing MeOH content and collecting 8 fractions (A).

Cytotoxic fractions A5, A6 and A8 were combined to yield 13.6 g and subjected to Sephadex LH-20 eluted with 95% aqueous EtOH. Four fractions (B) were collected. Among them, B3 (11 g) showed the highest level of cytotoxicity and was eluted with acetone/hexane (2:3) on the Si gel to give 9 fractions (C1-C9). The active fraction C8 (1.9 g) was rechromatographed on a RP-18 column, using MeOH/H$_2$O (9:1) as eluent to afford the active fraction D1 (120 mg), D2 (16mg), and D3 (21 mg).

Besides the activity-directed isolation of active components, other 10 known compounds were also isolated from fraction A2, A3, A4, A5, A6 and A7 by repeated chromatography on normal and reverse phase.

NMR Data of D1

^{1}H NMR δ ppm (CDCl$_3$, 200 MHz): δ 7.65 (1H, d, J = 15.7 Hz, H-3), 7.06 (2H, m, H-5, 9); 6.91 (1H, d, J = 8.0 Hz, H-8), 6.31 (1H, d, J = 15.7 Hz, H-2), 5.35 (1H, m, H-5') and δ 5.00 (1H, m, H-6'), 4.37 (2H, m, H-1'), 4.27 (1H, m, H-2'), 3.93, s, (3H, OCH$_3$-6), 3.78 (1H, m, H-4'), 3.47 (1H, m, H-3'), 2.01 (2H, m, H-7'), 0.88 (3H, t, J = 6.9 Hz, the terminal CH$_3$). ^{13}C NMR δ ppm (CDCl$_3$, 50 MHz): δ 167.2 (s, C-1), 148.2 (s, C-7), 146.8 (s, C-6), 146.0 (d, C-3), 129.9 (s, C-5', 6'), 126.7 (s, C-4), 123.4 (d, C-9), 114.8 (d, C-8), 114.5 (d, C-2), 109.5 (d, C-5), 73.4 (d, C- 4'), 72.2 (d, C-2'), 72.1 (d, C-3'), 66.0 (t, C-1'), 55.9 (q, OCH$_3$), 27.7 (t, C-7'), 14.1(q, the terminal CH$_3$).

Hydrolysis of Fraction D

The ferulate mixture (2-4 mg) was hydrolyzed in 90% MeOH (1 mL) containing 0.4 M KOH at room temperature for 2 hrs. The resulting fatty alcohols were extracted with EtOAc dried and evaporated.

Trimethylsilylated (TMS) Ether Derivatives

The hydrolyzed sample was derivatized to form the TMS ethers with bis-(trimethylsilyl)-trifluoroacetamide (200 μL) containing 1% trimethylchloro-silane. The reaction was completed in a Teflon-lined screw cap tube at 75°C for 30 minutes with a brief N$_2$ treatment prior to the reaction (*3*).

Determination of the Position of the Double Bond

The double bond position was determined using the dimethyl disulfide (DMDS) additive method (3). Briefly, an EtOAc solution (0.1 mL) of TMS ether derivatives was dried with a nitrogen stream, incubated with DMDS (0.2 mL) containing I$_2$ (15 mg/mL) at 35°C for 30 minutes. After reduction of excess I$_2$ with saturated NaHSO$_3$ aqueous solution, EtOAc (0.1 mL) was added into the mixture, and after brief centrifugation, a sample (1 μL) of the upper phase was injected into the GC-MS.

Activity Screening

All the fractions or extracts (20 mg) were dissolved in EtOH (10 mL) to make the concentration as 0.2%. 10 mL of the solution was added to 10 mL system with cancer cell lines. If cell growth was 100% inhibited, the concentration of this fraction was diluted to 0.0002% for 100% inhibition. If they did, these fractions were considered as active enough to be further purification.

Cytotoxicity Analysis of D1

PC-3 prostate tumor cells and MCF-7 breast tumor cells were obtained from the #18 American Type culture collection (ATCC, Rockville, MD). P388 mouse leukemia cells and P388 cells transformed with pHaMDR1/A retrovirus carrying an MDR1 cDNA were a gift from Dr. W. Hait (4). Cells were maintained at 37°C in an atmosphere of 5% CO_2 and grown in RPMI 1640 supplemented with 10% fetal bovine serum (FBS), 50 units penicillin and 50 μg/ml streptomycin. Cells were routinely checked and found to be free of contamination by mycoplasma. Cell viability was assessed by the tetrazolium dye method (MTT) as previously described (5). Cells were plated in 96 well plates and incubated with various agents for 72 hours. Absorbance was measured at 570 nm using a Dynatech micro plate reader.

Determination of the Scavenging Effect of D1 on DPPH Radicals

This method was adapted from that of Chen and Ho (6). Basically, the tested fraction (final concentrations were 10, 20, 30, 40, 50 μm, respectively), was added individually to an EtOH solution (2.5 mL) of DPPH radical (final concentration was 1.0×10^{-4} M). the mixtures were shaken vigorously and left to stand in the dark for 30 min. Thereafter, the absorbency for the sample was measured using a spectrophotometer (Milton Roy, Model 301) at 517 nm against blank samples without DPPH. The control sample was the DPPH solution without adding the tested sample. Each sample was duplicated in the test, and the values were averaged. The IC_{50} was obtained by extrapolation from linear regression analysis.

Results and Discussion

Components of Fresh Guggul Gum

As reported, the gum resin of *C. Wightii* is a rich source of steroids. Among them, (*Z*)-guggulsterone (1.6%) and (*E*)-guggulsterone (0.4%) have been found to be mainly responsible for the hypolipedmic activity, while the other components appear to exert a significant synergistic effect with regard to the lipid-lowering activity (*1*). In our study, the chemical investigation of fresh guggul led to the isolation and identification of 10 known compounds (Figure 1, except for cholsterol), including (*Z*)-guggulsterone (56 mg) and (*E*)-guggulsterone (17 mg).

Activity Directed Purification

The MTT assay in MCF-7 breast tumor cells was used to isolate the bioactive fraction from guggul gum. The EtOAc extract of guggul gum was subjected to bioassay directed fractionation and yielded three active fractions (A5, A6 and A8) on elution with a solvent mixture of $CHCl_3$/MeOH with increasing MeOH content. The combined fractions were rechromatographed on a Sephadex LH-20 column and eluted with 95% aqueous EtOH to yield an active fraction (B3). This fraction was rechromatographed on silica gel column eluted with acetone/hexane (2:3) to yield an active fraction (C8). Fraction C8 was chromatographed on a RP-18 column to yield three chemically homogeneous active fractions: D1, D2 and D3.

Fraction D1 exhibited a red fluorescent spot under $UV_{365 nm}$ light on silica gel F_{254} plates. In the ^1H NMR spectrum, a group of proton signals was assigned to the feruloyl moiety [δ 7.65, d, (J = 15.7 Hz, H-3), δ 7.06, m, (H-5, 9); δ 6.91, d, (J = 8.0 Hz, H-8), δ 6.31, d, (J = 15.7 Hz, H-2), and δ 3.93, s, (3H, OCH_3-6)] (*7*). The signals for the aliphatic methylene groups (δ 1.24), a terminal methyl (δ 0.88, t, J = 6.9 Hz), and a carbinyl methylene (δ 4.37, m) were assignable to those of a long chain alcoholic residue. Three-oxygen substituted methine protons (δ 3.47-4.27) indicated the presence of highly oxygenated alcohol. Comparison of the data of D1 with those for guggultetrol (Figure 2) (*2*), indicated that the only substantial difference was the presence of two olefinic proton signals (δ 5.35, m and δ 5.00, m) and two allylic proton signals (δ 2.01, m). Thus, D1 was suggested to be a unsaturated alkyl ferulate with three hydroxyl groups at C-2', C-3' and 4'. This proposal was supported by the ^1H-^1H COSY spectrum, which displayed correlations between H-1' and H-2', H-2' and H-3', and H-3' and H-4'.

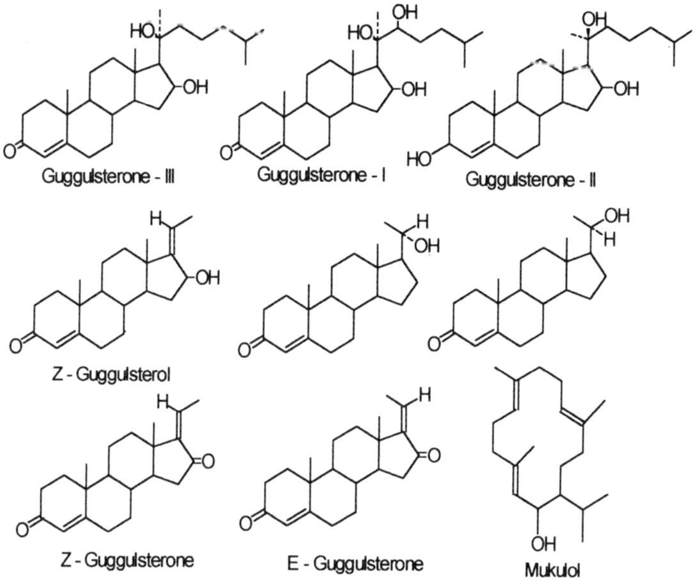

Figure 1. Structures of known compounds from gum guggul

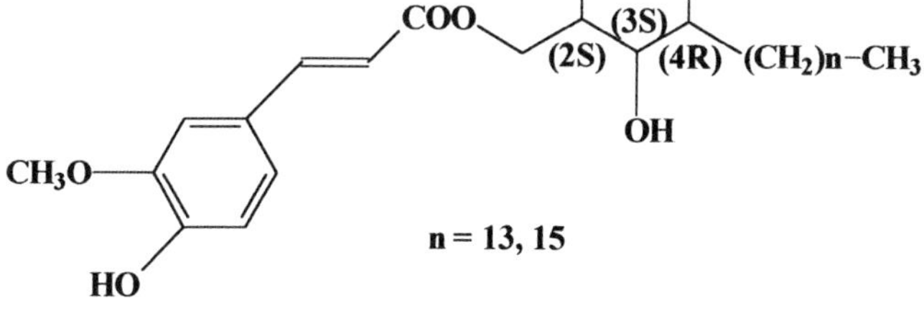

Figure 2. Structures of known ferulates

Alkaline hydrolysis of D1 yielded a mixture of two fatty alcohols. The DCI-MS data showed two $[M]^+$ molecular ions at m/z 562 and 576, respectively. In order to quantify and characterize the alcohol in D1, GC-MS and GC analyses were performed. Because of the substitution of multiple hydroxyl groups, the hydrolyzed alcohol mixture was first derivatized into TMS ethers. The double bond localization was carried out by GC-MS analysis of the corresponding bis(methylthio)-derivative vicinally substituted at the original double bond, which afforded an intense mass fragment arising from cleavage between the two carbons both substituted with a methylthio group (8). The resulting bis-(methylthio)-TMS ethers prepared from fraction D1 gave two major GC peaks, M1 and M2, which indicated that the predominant alcohols in fraction D1 were M1 (~81%) and M2 (~19%), respectively. In the GC-MS spectra, the peak M1 (R_f = 42.0 min) showed a fragment ion at m/z 299 (100%), which suggested that the localization of bis(methylthio) substitution at C-5 and C-6. Peak M2 (R_f = 49.1 min) showed a fragment ion at m/z 313 (100%), which also suggested localization of the bis(methylthio) substitution was at C-5 and C-6 (shown in Figure 3). The double bond localization was also supported by the NMR spectra. In the ^1H NMR spectrum, allylic proton signals indicated that only two hydrogen atoms were bonded with the double bond, while in the ^1H-^1H COSY spectrum, the allylic proton signal coupled with olefinic proton. In the HMBC spectrum, the double bond localization was further confirmed by correlations of H-7' to C-5' and C-6', and of H-5' to C-7'.

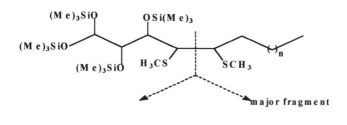

n = 16, major fragment: m/z 299
n = 17, major fragment: m/z 313

Figure 3. GC-MS fragment diagram of tetraols

The geometry of the double bonds in the alcohol was determined from the ^{13}C NMR data. The spectrum of the mixture exhibited an allylic carbon signal at δ 27.7, was reliably assigned to the methylene carbon adjacent to a *cis*-double bond in aliphatic chains. The corresponding allylic methylene carbons adjacent to a *trans*-bond are usually observed around δ 32.6 (*9*). This result suggested that the predominant alcohol in this fraction should have a *cis*-geometry. The absolute configuration of one the known ferulate (guggultetrol-18), was deduced as D-xylo (*2S*, *3S*, *4R*-configuration) by direct comparison with synthetic compounds (*2*). However, it was not possible to immediately determine the absolute stereochemistry of this fraction.

Combining the above information, the alcohols obtained by hydrolysis of D1 were concluded to be a mixture of (*z*)-5-tricosene-1,2,3,4-tetraol and (*z*)-5-tetracosene-1,2,3,4-tetraol (Figure 4).

To determine the structure of D2 and D3, the same method was used. NMR spectra suggested both D2 and D3 to be the unsaturated alkyl ferulate with three hydroxyl groups at C-2', C-3' and 4'. The geometry of the double bonds in the alcohol was also determined by ^{13}C NMR data, which proved the predominant alcohol in both fractions should have a *cis*-geometry. In GC spectrum, the resulted bis-(methylthio)-TMS ethers prepared from fraction D2 gave one major GC peak, which showed a fragment ion at m/z 327 (100%) in the GC-MS spectrum. This fact suggested that the localization of bis(methylthio) substitution at C-5 and C-6 with n = 18. Similarly, the resulted bis-(methylthio)-TMS ethers prepared from fraction D3 showed a fragment ion at m/z 341 (100%), which also suggested localization of the bis(methylthio) substitution was at C-5 and C-6 with n = 19 (Structures shown in Figure 4).

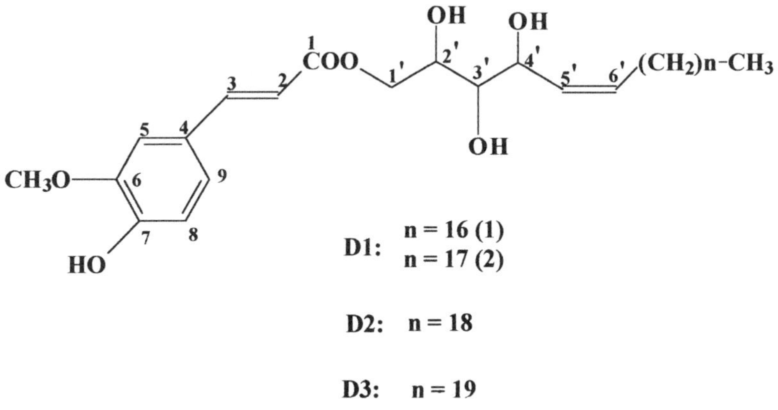

Figure 4. Structures of the ferulate D1, D2 and D3

Antioxidant analysis of D1

Five different concentrations of fraction D1 were made for DPPH test. Basically, the tested fraction (final concentrations were 10, 20, 30, 40, 50 μM, respectively), was added individually to an EtOH solution (2.5 mL) of DPPH radical (final concentration was 1.0×10^{-4} M). The mixtures were shaken vigorously and left to stand in the dark for 30 min. Thereafter, the absorbency for the sample was measured using a spectrophotometer at 517 nm against blank samples without DPPH. The IC_{50} was determined by extrapolation from linear regression analysis as 16.0 mcg/ml (28 μM) (Figure 5). It's known that gallic acid and catechin are essential to the antioxidant activity of tea (6), and their IC_{50} were reported as 10.6 μM and 18.6 μM, respectively (10). Therefore, the scavenging effect of fraction D1 on DPPH radicals was moderate compared with gallic acid and catechin.

Cytotoxic Activity of D1

More importantly, fraction D1 showed a significant cytotoxic activity. Fraction D1 decreased cell viability in MCF–7 (breast tumor cells) and PC-3 (prostate) tumor cells. In these tests, cells were maintained at 37°C in an atmosphere of 5% CO_2 and grown in RPMI 1640 supplemented with 10% fetal bovine serum (FBS), 50 units penicillin and 50 μg/mL streptomycin. Cells were routinely checked and found to be free of contamination by mycoplasma. Cell viability was assessed by the tetrazolium dye method (MTT). Cells were plated in 96 well plates and incubated with various agents for 72 hours. Absorbance was measured at 570 nm using a Dynatech micro plate reader. The IC_{50} in both cells were 14.3 mcg/ml (25 μM). To determine if fraction D1 could overcome a common mechanism of tumor resistance, P388 tumor cells transfected to over express the multidrug resistance pump P-glycoprotein were used. Fraction D1 decreased cell viability with IC_{50} < 25 μM in both transfected (P388/MDR) and parental cell lines This fact suggested that this agent might be able to overcome P-glycoprotein mediated drug resistance.

In summary, 10 known compounds, including the major hypolipidemic active component (Z)-guggulsterone and (E)-guggulsterone, were isolated from 'fresh' guggul gum. but, the amount of these steroids was much less than that found in aged sample. This fact may partly account for non-hypolipidemic activity of fresh sample. The cytotoxicity-directed fractionation has led to the identification of three mixtures from *Commiphora wightii* as bioactive compounds. These mixtures are a new class of naturally occurring lipids and shows strong cytotoxic activity and free radical scavenging activity.

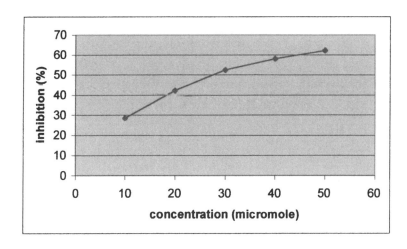

Figure 5. Scavenging effect of fraction D1 from gum guggul on DPPH radicals

References

1. Satyavati, G. V. *Economic and Medicinal Plant Research* **1991**, *5*, 47-82.
2. Kumar, V.; Dev, S. *Tetrahedron* **1987**, *43*, 5933-5948.
3. Wang, C. In *The roles of lipids in disease resistance and fruit ripening of tomato. Ph. D. Thesis*. **1998**. Rutgers University, New Brunswick, NJ.
4. Yang, J-M.; Goldenberg, S.; Gottesman, M. M.; Hait, W. N. *Cancer Research* **1994**, *54*, 730-737.
5. Scudiero, D. A.; Shoemaker, R. H.; Paull, K. D.; Monks, A.; Tierney, S.; Nofziger, T. H.; Currens, M. J., Seniff, D.; Boyd, M. E. *Cancer Research.* **1988**, *48*, 4827-4833.
6. Chen, C. W.; Ho, C.-T. *J. Food Lipids* **1995**, *2*, 35-46.
7. Katagiri, Y.; Mizutani, J.; Tahara, S. *Phytochemistry* **1997**, *46*, 347-352.
8. Francis, G. W.; Veland, K. *J. Chromatogr.* **1981**, *219*, 379-384.
9. Batchelor, J. G.; Cushley, R. J.; Prestegard, J. H. *J. Org. Chem.* **1974**, *39*, 1608-1705.
10. Chen, Y.; Wang, M.; Rosen, R. T.; Ho, C.-T. *J. Agric. Food Chem* **1999**, *47*, 2226-2228.

Chapter 21

Antioxidant Activity of Flavanols and Flavonoid Glycosides in Oolong Tea

Shengmin Sang[1], Nanqun Zhu[1], Shoei-Yn Lin-Shiau[2],
Jen-Kun Lin[3], and Chi-Tang Ho[1]

[1]Department of Food Science, Rutgers University, 65 Dudley Road, New
Brunswick, NJ 08901
[2]Institutes of Toxicology and [3]Biochemistry, College of Medicine,
National Taiwan University, Taipei, Taiwan

Fourteen flavanols and flavonoid glycosides were isolated
from the extracts of Oolong tea. Their structures were
determined by spectral methods (MS and NMR) and
compared with authentic samples. One of them is new to the
constituents of tea. DPPH free radical scavenging activity was
used to evaluate their antioxidant activity. The most active
compounds were found to be compounds **1-6**, compound **12**
and the mixture of compounds **13** and **14**.

The tea plant (*Camellia Sinensis*) is an evergreen tree belonging to the
family of Theaceae. Tea as a beverage has been consumed in many countries
for a very long time. Tea can be divided into three types-unfermented,
semifermented, and fermented, in terms of the degree of leaf fermentation
(Figure 1). Each type of tea has a distinct aroma, color, and flavor, which
appeal to our senses of taste, smell, and sight. Green tea, which is unfermented
tea, remains the most popular tea in the Asian countries such as China and
Japan. About 90% of the world's green tea is produced in China. Black tea is
fully fermented and accounts for approximately 70% of world's tea

293

consumption. The largest producer of black tea is India. China, Indonesia and Kenya are also major black tea producers. Oolong tea is partially fermented and accounts for less than 3% of world consumption and is produced both in the Mainland China as well as Taiwan.

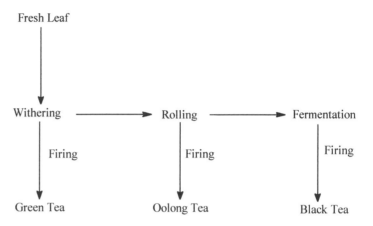

Figure 1. Tea manufacturing procedure

According to Chinese legends, the story of tea began in 2737 BC. Emperor Shen Nong discovered Tea. In 800 AD, the returning Buddhist priest Yeisei brought the first tea seeds to Japan. Later tea use was introduced into what is now known as Indonesia and from there through the Dutch colonials into Holland. It was also cultivated in India and subsequently imported to England, where it became popular. In the middle of the 17th century the English played a major role in merchandising and popularizing tea. In the United States of America, tea plays a dramatic role in the establishment of the USA. Most American knows the Boston tea party which is famous in the history of American Independence. United State of America was also the birthplace of iced tea and teabags.

Health Benefits of Tea

A large body of information has been published concerning the effects of tea and its major constituents on human health. A number of chronic diseases, such as cardiovascular disease and cancer, have been associated with excess production of reactive oxygen species and oxidative damage. Antioxidants have been identified as likely candidates responsible for the health benefits of many dietary plant materials. Tea has been proposed as a healthy beverage due to its powerful antioxidant activity. Although the antioxidant activity of catechins in teas has been well-recognized (1,2), the antioxidant activity of other phenolic compounds such as flavonols and flavonoid glycosides has not been extensively studied, especially for oolong tea. In this report, we have isolated flavonols and flavonoid glycoosides from oolong tea and studied their free radical scavenging activity.

Tea Polyphenols

Tea contains a wide range of polyphenols, such as catechins, flavonols and glycosides, flavones and glycosides, and phenolic acids and esters. In green tea, catechins are the predominant polyphenols. The major catechins in tea leaves are (-)-epigallocatechin 3-O-gallate (EGCG), (-)-epigallocatechin (EGC), (-)-epicatechin 3-O-gallate (ECG), (-)-epicatechin (EC), (+)-gallocatechin (GC) and (+)-catechin (C) (Figure 7). Epicatechins account for up to 15% (w/w) of the dry leaves; EGCG comprises 6-10% of the dry weight (3). These compounds contribute to the bitterness, astringency and sweet aftertaste of tea beverages (4). Moreover, two new catechin derivatives (Epigallocatechin-3-O-(3-O-methyl) gallate and Epigallocatechin-3-O-(4-O-methyl) gallate) with potent antiallergic activity were isolated from Taiwanese oolong tea (27). The characteristic color of black tea is generated during its manufacturing process. During this process, the colorless catechins are oxidized both enzymatically and chemically to give two major groups of pigments, theaflavins and thearubigins (5). The theaflavins are formed by oxidative coupling of the dihydroxybenzene and trihydroxybenzene rings of an appropriate pair of flavan-3-ols (e.g. epicatechin and epigallocatechin, respectively), resulting in a benzotropolone ring, which gives a yellow color. It is known that theaflavins make important contributions to the properties of black tea such as color (6), 'mouthfeel' (7) and extent of tea cream formation (8) and therefore studies of the structures of these pigments have long been of interest. Stereoisomers of theaflavin such as iso- and neotheaflavin have been identified (9,10) (Figure 2) and a number of closely related polyphenolic compounds, including theaflavic acids (10,11) and theaflagallins (12) have also been isolated from black tea. Recently, three novel minor polyphenol compounds, theaflavate B, isotheaflavin-3'-O-gallate and neotheaflavin-3-O-gallate, have been characterized in extracts from black tea (13). By contrast, thearubigins are extremely complex, heterogeneous mixture of pigments, and their structures are largely unknown (14). Thearubigins are predominant in black tea leaves (15-20% dry weight) and are believed to make the greatest contribution to taste, depth of color and body of a tea brew and therefore influencing the quality. Recently, a new polyphenolic pigment, named theacitrin A, has been isolated from the thearubigin fractions of an Assam black tea (15). Proanthocyanidins are colorless precursors of anthocyanidins (Figure 3). In tea a number of proanthocyanidins have been described by Japanese groups earlier (16-18), especially a series of B,B'-linked bisflavanoids, theasinensins A-G (Figure 3), isolated from the oolong tea (28). More recently, Lakenbrink and Engelhardt (19) reported two new proanthocyanidins in green tea. Contents of proanthocyanidins in green tea are roughly in the same range as the flavonol glycosides (20). In black tea the contents are much lower since there is a decrease during fermentation (21).

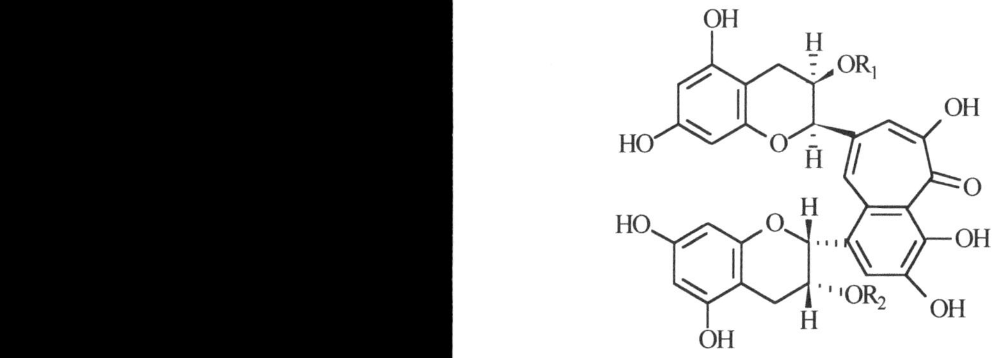

Theaflavin: $R_1=R_2=H$
Theaflavin-3-gallate: $R_1=H$; $R_2=galloyl$
Theaflavin-3'-gallate: $R_1=galloyl$; $R_2=H$
Theaflavin-3,3'-gallate: $R_1=R_2=galloyl$

Figure 2. Structures of the major theaflavins in tea

The flavonols and flavones in tea are present as aglycones (traces) and to a much higher extent as their glycosides. In tea leaves all the flavonol glycosides derive from the aglycones myricetin, quercetin and kaempferol, and the flavone glycosides derive from the aglycone apigenin (Figure 4). For the flavonol glycosides, the carbohydrate moieties, in most instances located at position 3 of the aglycone, consist of various combinations of glucose, galactose, and rhamnose and, in a single case, fructose. Mono-, di- and triglycosides have been observed. In contrast to flavonol glycosides, flavone glycosides in tea appear nearly exclusively as C-glycosides (22).

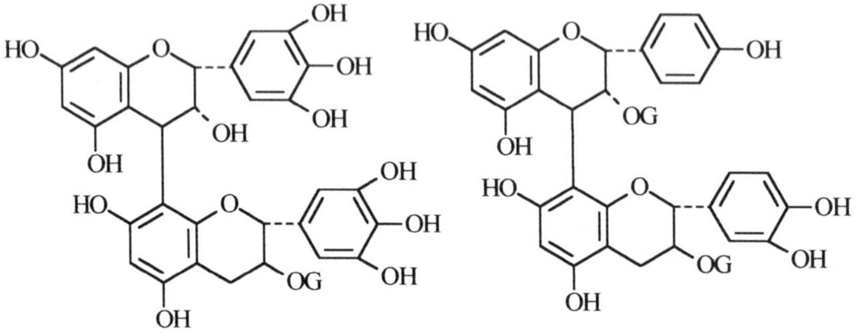

A,C'-Linked proanthocyanidin

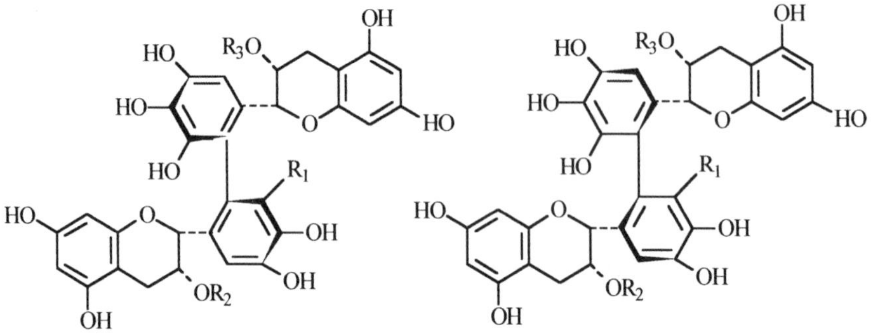

Theasinensin A: R₁=OH; R₂=R₃=G
Theasinensin B: R₁=OH; R₂=G; R₃=H
Theasinensin C: R₁=OH; R₂=R₃=H
Theasinensin F: R₁=H; R₂=R₃=G

Theasinensin D: R₁=OH; R₂=R₃=G
Theasinensin E: R₁=OH; R₂=R₃=H
Theasinensin G: R₁=H; R₂=R₃=G

Figure 3. Some structures of proanthocyanidins (G = gallate).

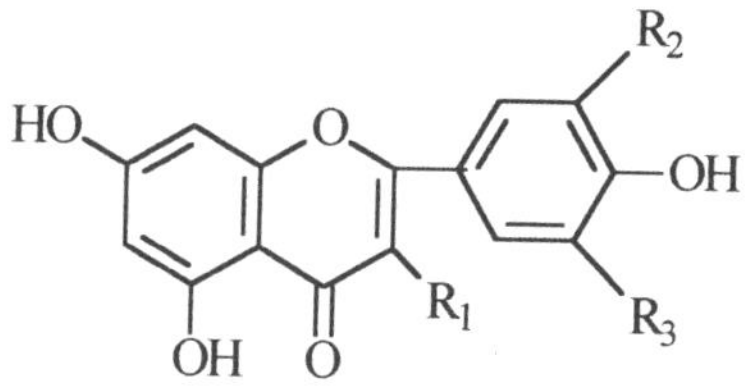

Kaempferol: R_1=OH, R_2=R_3=H
Quercetin : R_1=R_2=OH, R_3=H
Myricetin : R_1=R_2=R_2=OH
Apigenin   : R_1=R_2=R_3=OH

Figure 4. Structures of flavonol and flavone aglycones in tea.

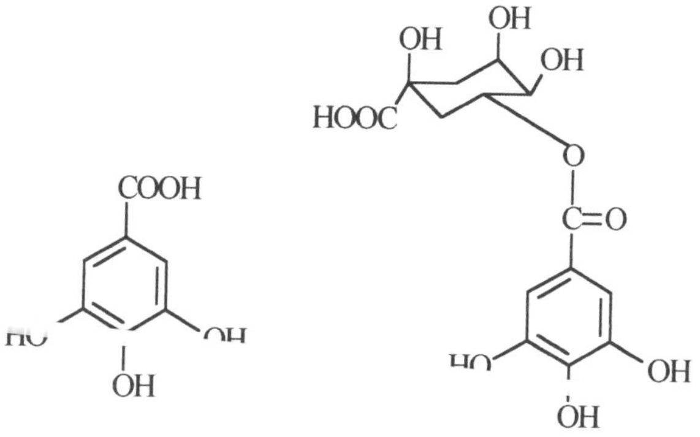

Gallic acid   Theogallin

Figure 5. Structures of gallic acid and theogallin.

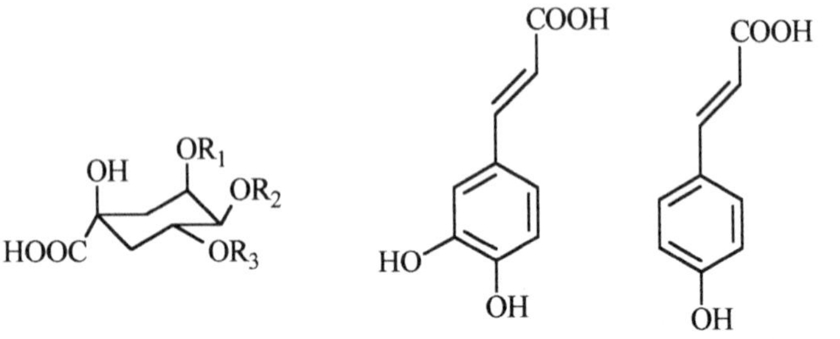

Derivatives of Quinic acid   Caffeic acid   p-coumaric acid

3-CQA   : $R_2=R_3=H$; $R_1=$Caffeic acid
4-CQA   : $R_1=R_3=H$; $R_2=$Caffeic acid
5-CQA   : $R_1=R_2=H$; $R_3=$Caffeic acid
3-CouQA: $R_2=R_3=H$; $R_1=$p-coumaric acid
4-CouQA: $R_1=R_3=H$; $R_2=$p-coumaric acid
5-CouQA: $R_1=R_2=H$; $R_3=$p-coumaric acid

Figure 6. Structures of caffeoyl- and p-coumaroylquinic acids.

The major phenolic acids and esters of tea constituents are gallic acid and its tea-specific ester with quinic acid (theogallin) (Figure 5) and hydroxycinnamoyl acid/quinic acid esters (Figure 6). Gallic acid is the most important phenolic acid in tea. The amount of gallic acid increases during the fermentation owing to its liberation from catechin gallates. Theogallin is a particularly interesting compound as it specifically appears in tea (23). Derivatives of hydroxycinnamic acids are widely distributed in the plant kingdom. Caffeoyl and *p*-coumarylquinic acids (CQAs and CouQAs) have been described in tea (24).

Material and Methods

Chemicals

Silica gel (130-270 mesh), Sephadex LH-20 (Sigma chemical Co., St. Louis, MO) and Lichroprep RP-18 column were used for column chromatography. All solvents used for chromatographic isolation were analytical grade and purchased from Fisher Scientific (Springfield, NJ).

General Procedures

^{1}H NMR and ^{13}C NMR spectra were obtained on a VXR-200 instrument (Varian Inc., Palo Alto, CA), operating at 200 and 50 MHz. Compounds were analyzed in CD$_3$OD with tetramethylsilane (TMS) as an internal standard. APCI MS (atmospheric pressure chemical ionization mass spectrometry) was obtained on a Fisons/VG Platform II mass spectrometer. Thin-layer chromatography was performed on TLC plates (250 μm thickness, 2-25 μm particle size, Sigma-Aldrich), with compounds visualized by spraying with 5% (v/v) H$_2$SO$_4$ in ethanol.

Extraction and Isolation Procedures

The dried leaves of oolong tea (1.5 kg) were extracted with hot water 3 times. After concentration, the extract was passed through a Diaion HP-20 column eluted with the water-ethanol solvent system. The adsorbed material was eluted with ethanol/water (60%) to give a brown residue (150 g). A part of this residue (30 g) was chromatographed on silica gel column, Sephadex LH-20 column and preparative TLC plate to afford 14 compounds. The concentrations of these compounds are shown in Table I.

Table I. Concentration of the Flavanols and Flavonoid Glycosides of Oolong Tea.

Compound	% Dry weight (mg/100g)
1	2.7
2	10
3	120
4	93
5	47
6	270
7	3.3
8	8.0
9	3.7
10	1.3
11	100
12	73
13+14	20

300

Results and Discussion

Structure determination of Isolated Compounds

To identify compounds isolated, we used ^{1}H-NMR, ^{13}C-NMR, CI-MS, and authentic samples. To make sure the structures were accurately assigned, the data of these compounds were compared with those in the literature (22,25). The structures of these compounds are shown in Figures 7 and 8. Among them, compound 9 is a novel compound isolated for the first time from tea leaves.

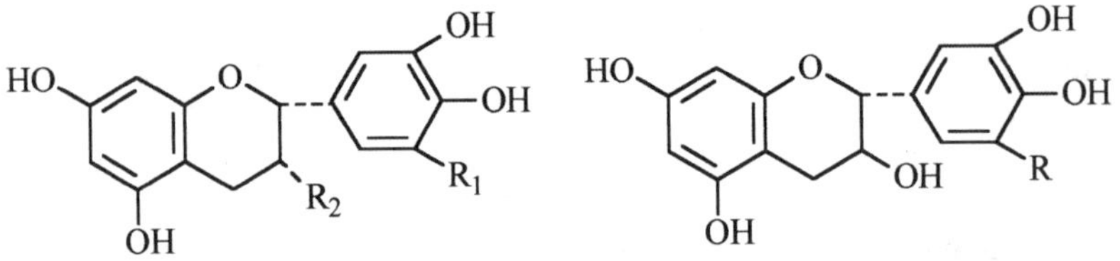

(1) EC   : R_1=R_2=H
(2) EGC : R_1=OH, R_2=H
(3) ECG : R_1=H, R_2=Gallate
(4) EGCG : R_1=OH, R_2=Gallate

(5) C : R=H
(6) GC : R=OH

Figure 7. Structures of compounds 1-6.

Determination of the Scavenging effect of Oolong Tea Flavanols and Flavonoid Glycosides on DPPH Radicals

This method was adapted from Chen and Ho (26). DPPH radicals were prepared in ethanol (1.0×10^{-4} M). This DPPH solution was mixed with different compounds (final concentration was 20 μM) and kept in a dark area for 0.5 h. The absorbance of the samples was measured on a spectrophotometer (Milton Roy, model 301) at 517 nm against a blank consisting of ethanol without DPPH. All tests were run in triplicate and mean values reported.

The scavenging effect of compounds **1-14** is shown in Table II. All compounds at a concentration of 20 μM exhibited scavenging activity compared to the control sample. Compounds **1-6**, compound **12** and the mixture of compound **13** and **14** showed strong DPPH radical scavenging activity.

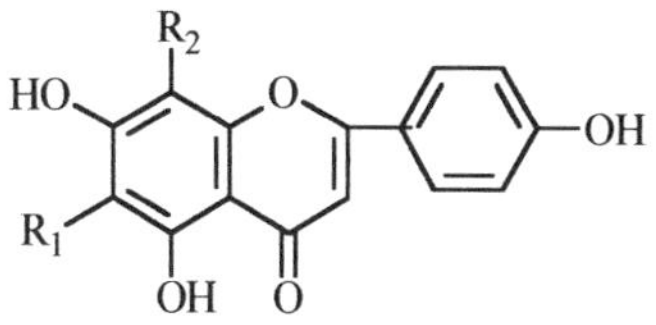

(7): R_1=Rha$\xrightarrow{1\rightarrow2}$ Glc R_2=H
(8): R_1=Ara R_2=Glc
(9): R_1=H R_2=Rha

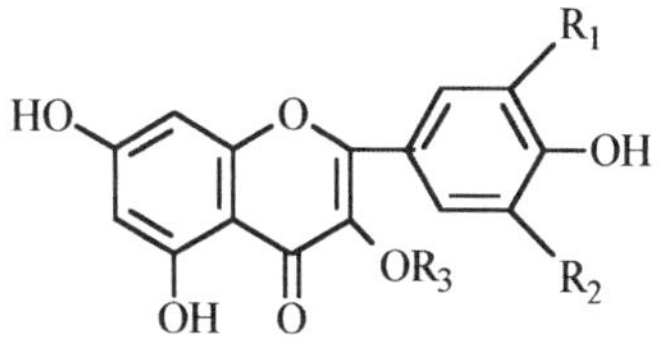

(10): R_1=R_2=H, R_3=Rha $\xrightarrow{1\rightarrow6}$ Glc

(11): R_1=R_2=H, R_3= $\begin{array}{c}\text{Rha} \xrightarrow{1\rightarrow6}\\ \text{Glc} \text{Glc}\\ 1\rightarrow2\end{array}$

(12): R_1=OH, R_2=H, R_3= $\begin{array}{c}\text{Rha} \xrightarrow{1\rightarrow6}\\ \text{Glc} \text{Glc}\\ 1\rightarrow2\end{array}$

(13): R_1=R_2=OH, R_3=Glc

(14): R_1=R_2=OH, R_3=Gal

Figure 8. Structures of compounds 7-14.

The DPPH radical-scavenging ability of catechins, as given in Table II, is in the order of EGCG>ECG>EGC>GC>C>EC, and the ability of the four aglycones is myricetin>quercetin>kaempferol>apigenin; and the ability of the two types glycoside is flavonol glycosides>flavone glycosides. By comparing the ability of the aglycones and the ability of the glycosides, it is clear that sugars of these glycosides do not play an important role in their activity

According to our study, the flavonol and flavone glycosides also play an important role in the antioxidant activity of tea. The structural requirements of DPPH radical inhibitors are: 1) requirement of the 3',4'-dihydroxy group in the B ring configuration; 2) enhancement in activity with the additional hydroxy group at 3 position or 5' of B ring. We also noticed that the fermentation procedure in oolong tea did not affect the integrity of flavonoid glycosides.

**Table II. Scavenging Effects of Compounds1-14 and Four Aglycons on
DPPH Radical**

Compound	Absorbance at 517 nm	Inhibition (%)
1	0.181	81.1
2	0.119	87.6
3	0.109	88.6
4	0.082	91.4
5	0.154	84.2
6	0.152	84.0
7	0.593	38.2
8	0.746	22.3
9	0.784	18.3
10	0.718	25.2
11	0.868	9.6
12	0.257	73.2
Mixtures of 13 and 14	0.147	84.7
Quercetin	0.207	78.4
Apigenin	0.796	17.1
Kaempferol	0.329	65.7
Myricetin	0.174	81.2
DPPH	0.960	---

References

1. Xie B.J.; Shi H.; Ho, C.-T. *Proceedings of the National Science Council, ROC Part B: Life Sciences.* **1993**, *17*, 77-84.
2. Chen C. W.; Ho C.T. *Journal of Food Lipids.* **1995**, *2*, 35-46.
3. Hara, Y.; Luo, S.J. *Food Reviews International*, **1995**, *11*, 435-456.
4. Hara, Y.; Luo, S.J. *Food Reviews International*, **1995**, *11*, 477-525.
5. Roberts, E.A.H.; Cartright, R.A. *J. Sci. Food Agric.* **1957**, *8*, 720.
6. Roberts, E.A.H. *J. Sci. Food Agric.* **1959**, *9*, 381.
7. Millin, D.J.; Crispin, D.J. *J. Agric. Food Chem.* **1969**, *17*, 717.
8. Powell, C.; Clifford, M.N. *J. Sci. Food Agric.* **1992**, *63*, 77.
9. Collier, P.D.; Bryce, T. *Tetrahedron*, **1973**, *29*, 125.
10. Robertson, A. *Tea: Cultivation to consumption.* Eds. K. Wilson and M. Clifford. **1992**, p 555.
11. Bryce, T.; Collier, P.D. *Tetrahedron Lett.* **1972**, 463.
12. Nishioka, I.; Nonaka, G.I. *Chem. Pharm. Bull.* **1986**, *34*, 61.

13. Lewis, J.R.; Davis, A. L. *Phytochemistry*, **1999**, *49*, 2511-2519.
14. Opie, S.C.; Robertson, A. *J. Sci. Food Agric.* **1990**, *50*, 547.
15. Davis, A.L.; Lewis, J.R. *Phytochemistry*, **1997**, *46*, 1397-1402.
16. Nonaka, G.I.; Kawahara, O. *Chem. Pharm. Bull.* **1983**, *31*, 3906-3914.
17. Nonaka, G.I.; Sakai, R. *Phytochemistry*, **1984**, *23*, 1753-1755.
18. Hashimoto, F.; Nonaka, G.I. *Chem. Pharm. Bull.* **1989**, *37*, 3255-3263.
19. Lakenbrink, C.; Engelhardt, U.H. *J. Agric. Food Chem.* **1999**, *47*, 4621-4624.
20. Degenhardt, A.; Engelhardt, U.H. *J. Agric. Food Chem.* **2000**, *48*, 3425-3430.
21. Hashimoto, F.; Nonaka, G. *Chem. Pharm. Bull.* **1992**, *40*, 2379-2383.
22. Finger A.; Kuhr S. *Journal of Chromatography.* **1992**, *624*, 293-315.
23. Roberts, E.A.H. *J. Sci. Food Agric.* **1958**, *9*, 701.
24. Roberts, E.A.H. *J. Sci. Food Agric.* **1958**, *9*, 212.
25. Yu D.Q.; Yang J.S. *Handbook of Analytical Chemistry: NMR Spectrum Analysis; Beijing, P.R. China,* **1989**; Vol.5; pp: 735-758.
26. Chen, J.H.; Ho, C.T. *J. Agric. Food Chem.* **1997**, *45*, 2374-2378.
27. Sano, M.; Suzuki, M. *J. Agric. Food Chem.* **1999**, *47*, 1906-1910,
28. Hashimoto, F.; Nonaka, G. *Chem. Pharm. Bull.* **1988**, *36*, 1676-1684.

INDEXES

Author Index

Subject Index

A

research needs, 60
reversed-phase HPLC, 51, 53
reversible structural transformations
with pH, 43, 45*f*
sample detection method, 53, 56
sample preparation, 53
total pigment by pH differential
method, 47, 49
total pigment content equation, 47,
49
Anthraquinones
analysis of noni, 135
glycosides from flowers of *Morinda
citrifolia*, 137*f*
seeds and heartwood of *M. citrifolia*,
136*f*
See also Morinda citrifolia
Anti-fatigue. *See* Maca
Antioxidants
analysis of fraction from guggul
gum, 290
search for, from spices and herbs,
230–231
stabilizing marine oils, 81, 82*t*
See also Ginger family; Oolong tea
Anti-platelet aggregating effects,
garlic products, 69
Antitumor promoting effects,
components of noni, 148–149
Apigenin, *Achillea millefolium*, 33
Apocynin D, *Apocynum venetum*,
35*f*
Apocynum venetum
flavan-3-ols, 33*f,* 34*f*
kaempferol and kaempferol
glycosides, 25*f*
quercetin and quercetin glycosides,
27*f*
Asperuloside, acetyl derivatives in
Morinda citrifolia, 135, 138*f*
Atmospheric pressure chemical
ionization (APCI), mass spectrum
of peaks from *Polygonatum
odoratum*, 274*f*

B

Benign prostatic hyperplasia (BPH),
saw palmetto, 118
Berberine
alkaloid in *Hydrastis canadensis*, 200
calibration curve, 208*f*
existence in other botanicals, 200–
201, 207–208
ion-pairing HPLC for separation of,
and hydrastine, 203
linear dynamic range, 207
percent recovery, 203, 207
structure, 201*f*
See also Goldenseal
Biflavones, *Ginkgo biloba* and
Hypericum perforatum, 30*f*
Bioactive ferulates. *See* Guggul
Bioactive peptides, marine
nutraceuticals, 85
Bioassay-directed fractionation
bcl-2 protein, 270
See also Polygonatum odoratum
Bioassays, extracts from grape seeds,
175, 177
Bioavailability, garlic compounds, 68–
69
Black cohosh
alcoholic extract of *Cimicifuga
foetida*, 97
alcoholic extract of *C. racemosa*, 96
alternative to estrogen replacement
therapy (ERT), 98
assay, 105, 108–109
atmospheric pressure chemical
ionization (APCI), 91–92
calibration, 110
Cimicifuga species, 91, 92
Cimicifuga species identification by
LC/MS, 97–98
commercially available products by
selected ion chromatogram (SIC),
98, 99*f*
commercial preparations, 91

Highlights from ACS Books

Desk Reference of Functional Polymers: Syntheses and Applications
Reza Arshady, Editor
832 pages, clothbound, ISBN 0–8412–3469–8

Chemical Engineering for Chemists
Richard G. Griskey
352 pages, clothbound, ISBN 0–8412–2215–0

Controlled Drug Delivery: Challenges and Strategies
Kinam Park, Editor
720 pages, clothbound, ISBN 0–8412–3470–1

A Practical Guide to Combinatorial Chemistry
Anthony W. Czarnik and Sheila H. DeWitt
462 pages, clothbound, ISBN 0–8412–3485–X

Chiral Separations: Applications and Technology
Satinder Ahuja, Editor
368 pages, clothbound, ISBN 0–8412–3407–8

Molecular Diversity and Combinatorial Chemistry: Libraries and Drug Discovery
Irwin M. Chaiken and Kim D. Janda, Editors
336 pages, clothbound, ISBN 0–8412–3450–7

A Lifetime of Synergy with Theory and Experiment
Andrew Streitwieser, Jr.
320 pages, clothbound, ISBN 0–8412–1836–6

For further information contact:
Order Department
Oxford University Press
2001 Evans Road
Cary, NC 27513
Phone: 1-800-445-9714 or 919-677-0977
Fax: 919-677-1303

More Best Sellers from ACS Books

Microwave-Enhanced Chemistry: Fundamentals, Sample Preparation, and Applications
Edited by H. M. (Skip) Kingston and Stephen J. Haswell
800 pp; clothbound ISBN 0–8412–3375–6

Designing Bioactive Molecules: Three-Dimensional Techniques and Applications
Edited by Yvonne Connolly Martin and Peter Willett
352 pp; clothbound ISBN 0–8412–3490–6

Principles of Environmental Toxicology, Second Edition
By Sigmund F. Zakrzewski
352 pp; clothbound ISBN 0–8412–3380–2

Controlled Radical Polymerization
Edited by Krzysztof Matyjaszewski
484 pp; clothbound ISBN 0–8412–3545–7

The Chemistry of Mind-Altering Drugs: History, Pharmacology, and Cultural Context
By Daniel M. Perrine
500 pp; casebound ISBN 0–8412–3253–9

Computational Thermochemistry: Prediction and Estimation of Molecular Thermodynamics
Edited by Karl K. Irikura and David J. Frurip
480 pp; clothbound ISBN 0–8412–3533–3

Organic Coatings for Corrosion Control
Edited by Gordon P. Bierwagen
468 pp; clothbound ISBN 0–8412–3549–X

Polymers in Sensors: Theory and Practice
Edited by Naim Akmal and Arthur M. Usmani
320 pp; clothbound ISBN 0–8412–3550–3

Phytomedicines of Europe: Chemistry and Biological Activity
Edited by Larry D. Lawson and Rudolph Bauer
336 pp; clothbound ISBN 0–8412–3559–7

For further information contact:
Order Department
Oxford University Press
2001 Evans Road
Cary, NC 27513
Phone: 1-800-445-9714 or 919-677-0977

Bestsellers from ACS Books

The ACS Style Guide: A Manual for Authors and Editors (2nd Edition)
Edited by Janet S. Dodd
470 pp; clothbound ISBN 0–8412–3461–2; paperback ISBN 0–8412–3462–0

Writing the Laboratory Notebook
By Howard M. Kanare
145 pp; clothbound ISBN 0–8412–0906–5; paperback ISBN 0–8412–0933–2

Career Transitions for Chemists
By Dorothy P. Rodmann, Donald D. Bly, Frederick H. Owens, and Anne-Claire Anderson
240 pp; clothbound ISBN 0–8412–3052–8; paperback ISBN 0–8412–3038–2

Chemical Activities (student and teacher editions)
By Christie L. Borgford and Lee R. Summerlin
330 pp; spiralbound ISBN 0–8412–1417–4; teacher edition, ISBN 0–8412–1416–6

Chemical Demonstrations: A Sourcebook for Teachers, Volumes 1 and 2, Second Edition
Volume 1 by Lee R. Summerlin and James L. Ealy, Jr.
198 pp; spiralbound ISBN 0–8412–1481–6
Volume 2 by Lee R. Summerlin, Christie L. Borgford, and Julie B. Ealy
234 pp; spiralbound ISBN 0–8412–1535–9

The Internet: A Guide for Chemists
Edited by Steven M. Bachrach
360 pp; clothbound ISBN 0–8412–3223–7; paperback ISBN 0–8412–3224–5

Laboratory Waste Management: A Guidebook
ACS Task Force on Laboratory Waste Management
250 pp; clothbound ISBN 0–8412–2735–7; paperback ISBN 0–8412–2849–3

Good Laboratory Practice Standards: Applications for Field and Laboratory Studies
Edited by Willa Y. Garner, Maureen S. Barge, and James P. Ussary
571 pp; clothbound ISBN 0–8412–2192–8

For further information contact:
Order Department
Oxford University Press
2001 Evans Road
Cary, NC 27513
Phone: 1-800-445-9714 or 919-677-0977